23rd International Conference on Computational Statistics (COMPSTAT 2018) and CRoNoS Summer Course and Satellite Workshop on Functional Data Analysis (CRoNoS 2018)

Iasi, Romania
28 August - 2 September 2018

ISBN: 978-1-5108-7265-3

Printed from e-media with permission by:

Curran Associates, Inc.
57 Morehouse Lane
Red Hook, NY 12571

Printed by Curran Associates, Inc. (2018)

For permission requests, please contact International Statistical Institute (ISI) at the address below.

International Statistical Institute (ISI)
P.O. Box 24070
2490 AB The Hague
The Netherlands

Phone: 31-70-3375737
Fax: 31-70-3860025

www.isi-web.org

Additional copies of this publication are available from:

Curran Associates, Inc.
57 Morehouse Lane
Red Hook, NY 12571 USA
Phone: 845-758-0400
Fax: 845-758-2633
Email: curran@proceedings.com
Web: www.proceedings.com

PROGRAMME AND ABSTRACTS

23rd International Conference on

Computational Statistics (COMPSTAT 2018)

http://www.compstat2018.org

Unirea Hotel, Iasi, Romania
28-31 August 2018

2018 CRoNoS Summer Course and Satellite Workshop on

Functional Data Analysis (CRoNoS FDA 2018)

http://www.compstat2018.org/CRoNoS_SummerCourse.php
http://www.compstat2018.org/SatelliteWorkshop.php

Alexandru Ioan Cuza University of Iasi, Iasi, Romania
31 August -2 September 2018

ALEXANDRU IOAN CUZA UNIVERSITY of IAŞI

ERCIM WG on Computational and Methodological Statistics
http://www.CMStatistics.org

COMPSTAT 2018 Scientific Program Committee:

Ex-officio:
COMPSTAT 2018 organiser and chairperson of the SPC: Cristian Gatu.
Past COMPSTAT organiser: Ana Colubi.
Next COMPSTAT organiser: Alessandra Luati and Maria Brigida Ferraro.
Incoming IASC-ERS Chairman: Peter Winker.

Members:
John Hinde, Karel Hron, Ivan Kojadinovic, Erricos J. Kontoghiorghes, Christophe Ley, Alfio Marazzi, Laura Sangalli.

Consultative Members:
Representative of the IFCS: Sugnet Lubbe.
Representative of the ARS of IASC: Jaeyong Lee.
Representative of the LARS of IASC: Paulo Canas Rodrigues.
Representative of CMStatistics: Peter Buhlmann.

Local Organizing Committee:
Mihaela Breaban, Mircea Ioan Cosbuc, Lucian Gadioi, Gabriela Mesnita, Daniela Zaharie.

CRoNoS FDA 2018 Scientific Program Committee:

Members:
Ana M. Aguilera, Enea Bongiorno, Frederic Ferraty, Gil Gonzlez, Alois Kneip, Stefan Van Aelst.

Organizers:
Ana Colubi and Cristian Gatu.

Dear Colleagues and Friends,

We wish to warmly welcome you to Iasi, for the 23rd International Conference on Computational Statistics (COMPSTAT 2018) and the CRoNoS Summer Course and Satellite Workshop on Functional Data Analysis (CRoNoS FDA 2018). These events are locally organized by members of the Alexandru Ioan Cuza University of Iasi assisted by renown international researchers. The COMPSTAT is an initiative of the European Regional Section of the International Association for Statistical Computing (IASC-ERS), a society of the International Statistical Institute (ISI). COMPSTAT is one of the most prestigious world conferences in Computational Statistics, regularly attracting hundreds of researchers and practitioners.

The first COMPSTAT conference took place in Vienna in 1974, and the last two editions took place in Geneva in 2014 and Oviedo in 2016. It has gained a reputation as an ideal forum for presenting top quality theoretical and applied work, promoting interdisciplinary research and establishing contacts amongst researchers with common interests.

Keynote lectures are addressed by Prof. Peter Rousseeuw, KU Leuven, Belgium, Prof. Steven Gilmour, King's College London, UK, and Prof. Alessandra Luati, University of Bologna, Italy.

From more than 300 submissions received for COMPSTAT, 260 have been retained for presentation in the conference. The conference programme has 22 contributed sessions, 10 invited sessions, 3 keynote talks, 28 organized sessions and 2 tutorials. There are approximately 300 participants. The CRoNoS Summer Course and Satellite Workshop have about 65 participants, 37 talks and 20 hours of lectures.

The organization would like to thank the authors, referees and all participants of COMPSTAT 2018 who contributed to the success of the conference. Our gratitude to sponsors, scientific programme committee, session organizers, local hosts, the city of Iasi, and many volunteers who have contributed substantially to the conference. We acknowledge their work and support.

The COMPSTAT 2018 organizers invite you to the next edition of the COMPSTAT which will take place in Bologna, Italy, 25-28 August 2020. We wish the best success to Alessandra Luati and Maria Brigida Ferraro, Chairs of the 24th COMPSTAT.

Ana Colubi and Cristian Gatu.

SCHEDULE

COMPSTAT 2018

2018-08-28

- Opening, 09:25 - 09:40
- A - Keynote, COMPSTAT2018, 09:40 - 10:30
- Coffee Break, 10:30 - 11:00
- B, COMPSTAT2018, 11:00 - 12:30
- Lunch Break, 12:30 - 14:15
- C, COMPSTAT2018, 14:15 - 15:45
- Coffee Break, 15:45 - 16:15
- D, COMPSTAT2018, 16:15 - 17:45
- Welcome Reception, 19:00 - 20:30

2018-08-29

- E, COMPSTAT2018, 09:00 - 10:30
- Coffee Break, 10:30 - 11:00
- F, COMPSTAT2018, 11:00 - 12:30
- Lunch Break, 12:30 - 14:15
- G, COMPSTAT2018, 14:15 - 16:15
- Guided Visit, 17:00 - 19:00

2018-08-30

- H, COMPSTAT2018, 09:00 - 11:00
- Coffee Break, 11:00 - 11:30
- I - Keynote, COMPSTAT2018, 11:30 - 12:20
- Lunch Break, 12:20 - 14:15
- J, COMPSTAT2018, 14:15 - 15:45
- Coffee Break, 15:45 - 16:15
- K, COMPSTAT2018, 16:15 - 17:45
- Conference Dinner, 20:30 - 23:00

2018-08-31

- L, COMPSTAT2018, 09:00 - 10:30
- Coffee Break, 10:30 - 11:00
- M, COMPSTAT2018, 11:00 - 12:00
- N - Keynote, COMPSTAT2018, 12:10 - 13:00
- Closing, 13:00 - 13:15
- Lunch Break, 13:00 - 14:30
- O, COMPSTAT2018, 14:30 - 16:30

SCHEDULE

CRoNoS FDA 2018

2018-08-31	2018-09-01	2018-09-02
	F 08:30 - 09:30	L 08:30 - 09:30
A 09:00 - 10:30	G - Keynote 09:40 - 10:30	M - Keynote 09:40 - 10:30
Coffee Break 10:30 - 11:00	Coffee Break 10:30 - 11:00	Coffee Break 10:30 - 11:00
B 11:00 - 12:00	H 11:00 - 13:00	N 11:00 - 12:30
	Lunch Break 13:00 - 14:30	Lunch Break 12:30 - 14:00
C 14:30 - 16:30	I 14:30 - 16:00	O 14:00 - 15:30
	Coffee Break 16:00 - 16:30	Coffee Break 15:30 - 16:00
Coffee Break 16:30 - 17:00	J 16:30 - 18:00	P 16:00 - 19:00
D 17:00 - 18:00		
E 18:00 - 19:00	K 18:00 - 19:00	
Welcome Reception 19:15 - 20:45		
	Workshop and Course Dinner 20:00 - 22:30	

TUTORIALS, SUMMER COURSE, MEETINGS AND SOCIAL EVENTS

TUTORIALS - COMPSTAT 2018

The first tutorial will take place at room Vega Hall, Unirea Hotel (see maps at pages IX and XI) and in parallel with the invited, organized and contributed sessions. It is given by Frederic Ferraty (*Functional data and nonparametric modelling: Theoretical/methodological/practical aspects*), Friday 31.08.2018, 09:00 - 10:30. The second tutorial will take place at Room C3, 2nd Floor of the Faculty of Computer Sciences, Alexandru Ioan Cuza University of Iasi (see maps at pages IX and XI). It is given by Manuel Escabias (*An online application for functional data analysis based on R*), Friday 31.8.2018, 14:30 - 16:30

SUMMER COURSE - CRoNoS FDA 2018

The summer course will take place at Rooms C3 and C413, 2nd Floor of the Faculty of Computer Sciences, Alexandru Ioan Cuza University of Iasi (see maps at pages IX and XI) and in parallel with the sessions of the Satellite CRoNoS Workshop. The course is given by Manuel Escabias, Manuel Febrero and Fang Yao.

SPECIAL MEETINGS by invitation to group members

- IASC Executive Committee meeting, *Room: Mezanin Veranda (entrance through the Mezanin room)*, Tuesday 28th August 2018, 12:35-14:10.
- ERS BoD Meeting, *Room: Mezanin Veranda (entrance through the Mezanin room)*, Wednesday 29th August 2018, 12:40-14:00.
- IASC and ERS General Assembly, *Room: Mezanin Lounge*, Thursday 30th August 2018, 17:50-18:50.

SOCIAL EVENTS - COMPSTAT 2018

- *The coffee breaks* will take place in front of the lecture rooms at the ground and the first floor of the Unirea Hotel (see maps at pages IX and XI). You must have your conference badge in order to attend the coffee breaks.
- *Set menu Lunches* will be arranged at the Grand Hotel Traian on 28th, 29th, 30th and 31st of August 2018 (see map at page IX). The lunches are optional and registration is required. Information about the purchased lunch tickets is embedded in the QR code on the conference badge. You must have your conference badge in order to attend the lunch each day. People not registered for lunch can buy lunch at restaurants and cafes in close walking distance to the conference venue.
- *Welcome Reception, Tuesday 28th of August, 19:00-20:30.* The Welcome Reception will take place at Roznovanu Palace (The City Hall), 11th, Stefan cel Mare si Sfant Blvd., Iasi (see map at page IX), and is open to all registrants (for free) who had preregistered and accompanying persons who have purchased a reception ticket. Participants must bring their conference badge in order to attend the reception. Preregistration is required due to health and safety reasons.
- *Botanical Garden Guided Visit, Wednesday 29th of August 2018, 17:00-19:00.* The Iasi Botanical Garden was established in the 1856 and is maintained by the Alexandru Ioan Cuza University of Iasi. It is the oldest and largest botanical garden in Romania. A guided visit for the participants to the conference has been arranged. Participants should be at the gates of the Botanical Garden, 7-9 Dumbrava Rosie Str., Iasi at 17:00 sharp. It is about 3.5 km from the Conference venue. You can get there by tram lines 1, 8, 9 or 13 getting in at stop "Piata Unirii", direction "Copou", and step out at the 5th stop, "Stadion". The Botanical Garden is on your left, behind the "Exhibition Park" (see map at page X). Participants must bring their conference badge in order to attend the visit. The event is free.
- *Conference Dinner, Thursday 30th August 2018, 20:30-23:00.* The conference dinner will take place at the Panoramic Restaurant of the Unirea Hotel. It is optional and registration is required. Information about the purchased conference dinner ticket is embedded in the QR code on the conference badge. Participants must bring their conference badge in order to attend the conference dinner.

SOCIAL EVENTS - CRoNoS FDA 2018

- *The coffee breaks* will take place in front of the room C3, 2nd Floor of the Faculty of Compute Science (see maps at pages IX and XI). You must have your CRoNoS FDA 2018 badge in order to attend the coffee breaks.
- *Welcome Reception, Friday 31st of August, 19:15-20:45.* The Welcome Reception will take place at the ground lobby of the R building of Alexandru Ioan Cuza Univeristy of Iasi, 28th Lapusneanu Str., Iasi (see map at page IX). It is free for all registrants. Summer course and satellite workshop registrants must bring their badge in order to attend the reception.
- *Summer Course and Satellite Workshop Dinner, Saturday 1st of September 2018, 20:00-22:30.* The CRoNoS FDA 2018 dinner is optional and registration is required. It will take place at "Casa Pogor" Restaurant Str. Pogor, n. 4, Iasi (see map at page IX). CRoNoS FDA 2018 registrants must bring their conference badge in order to attend the dinner.

Address of venue (see maps at page IX)

The Conference venue is the Unirea Hotel, at Piata Unirii 5, RO-700056 Iasi, Romania. The satellite events CRoNoS FDA venue is the Faculty of Computer Science, Alexandru I. Cuza University of Iasi, General Berthelot Str. 16, RO-700483 Iasi, Romania.

Registration

The registration will be open during the days of the conference from 08:40 to 17:00 and will take place at the lobby of the Unirea Hotel (see maps at pages IX and XI). On Friday 31st of August at 13:00 it will be moved to the area in front of Room C3, 2nd Floor of the Faculty of Computer Science for the satellite event (see maps at pages IX and XI).

Lecture rooms

The paper presentations will take place at the ground floor and the first floor of the Unirea Hotel. The location of the rooms can be checked at page XI. Rooms Cuza Hall/Cuza Center and Cocktail are in the ground floor. Rooms Vega, Clio, Orion and Mezanin are in the first floor. The opening, keynote and closing talks of the COMPSTAT 2018 conference will take place at the Cuza Center. The poster presentations will take place in the Cocktail Hall located in the ground floor.

Presentation instructions

The lecture rooms will be equipped with a mini-PC and a computer projector. The session chairs should obtain copies of the talks on a USB stick before the session starts (use the lecture room as the meeting place), or obtain the talks by email prior to the start of the conference. Presenters must provide to the session chair with the files for the presentation in PDF (Acrobat) format on a USB memory stick. This must be done ten minutes before each session. The session chairs are kindly requested to have a laptop for backup. IT technicians will be available during the conference and should be contacted in case of problems. The posters should be displayed only during their assigned session. The authors will be responsible for placing the posters in the poster panel displays and removing them after the session. The maximum size of the poster is A0.

Internet

Throughout the venue there will be wireless Internet connection available at the halls. Wifi information will be displayed by the registration desk.

Map of the COMPSTAT venue and Social Events

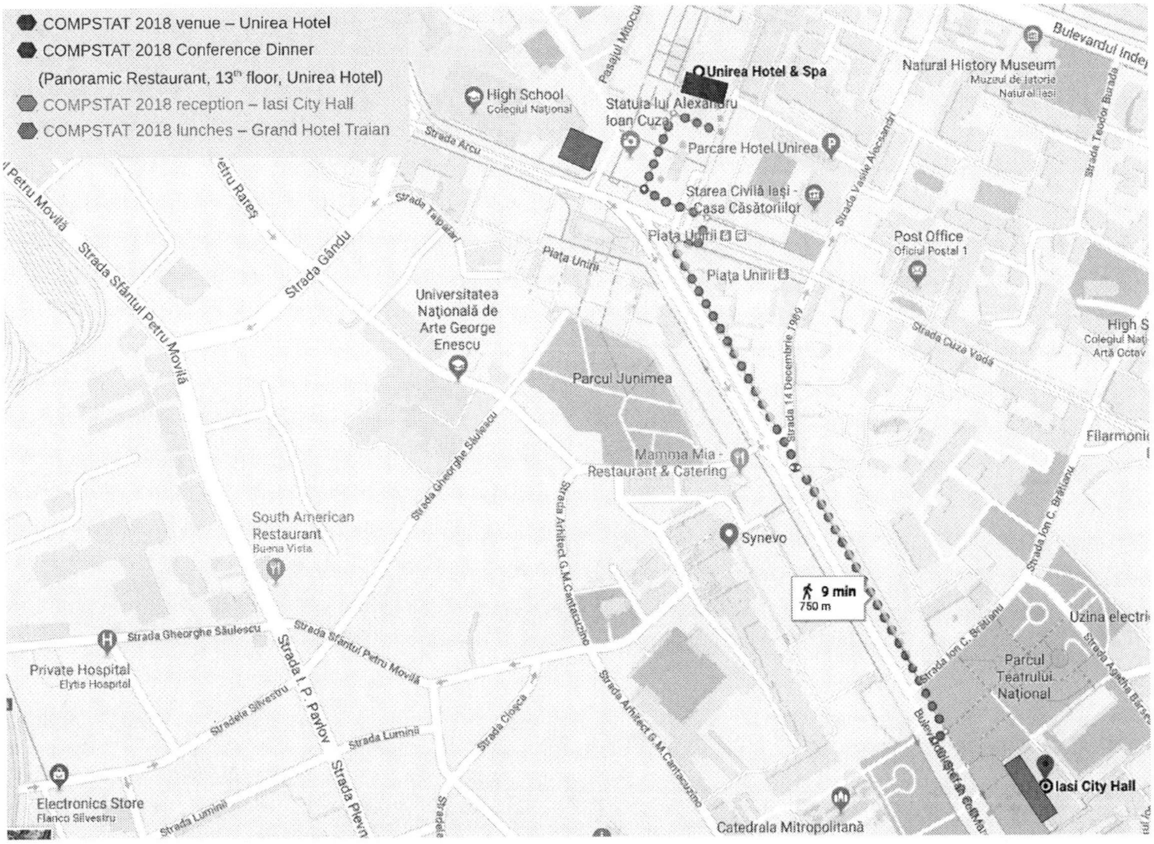

Map of the CRoNoS FDA venue and Social Events

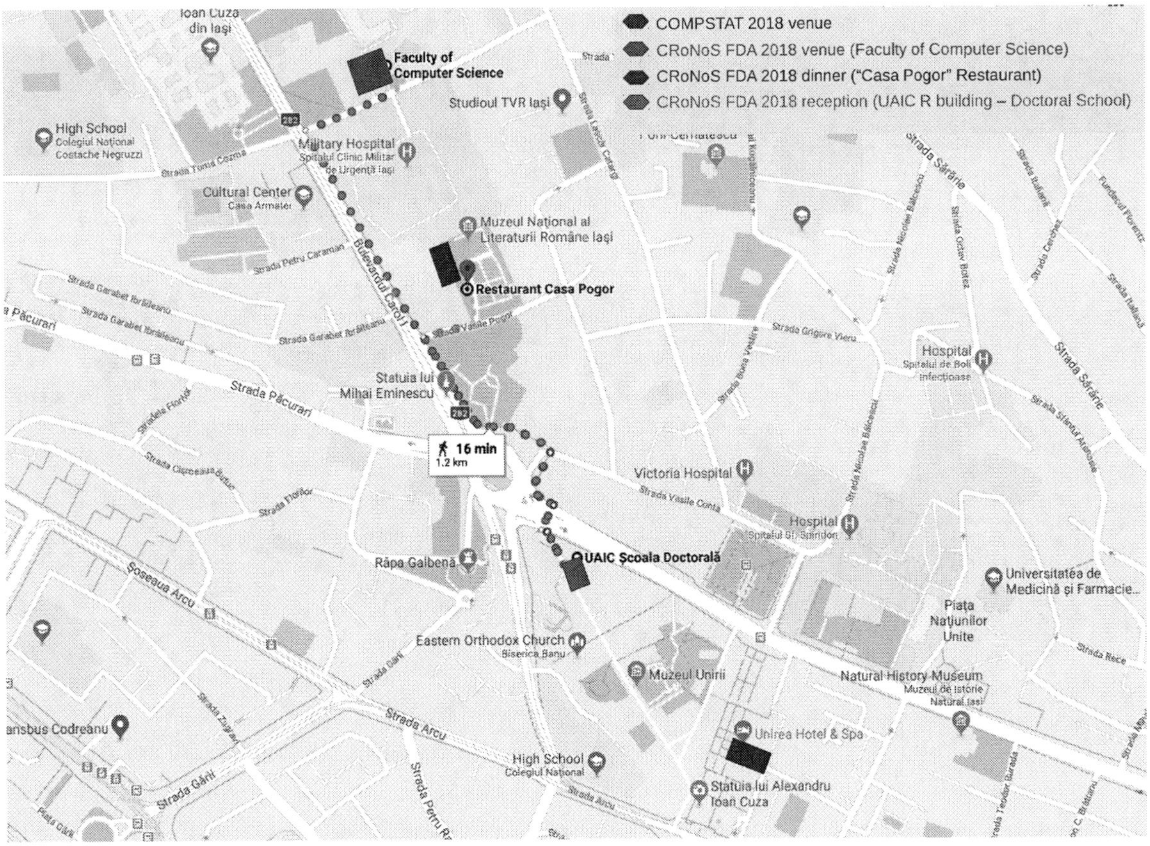

Map for the guided visit

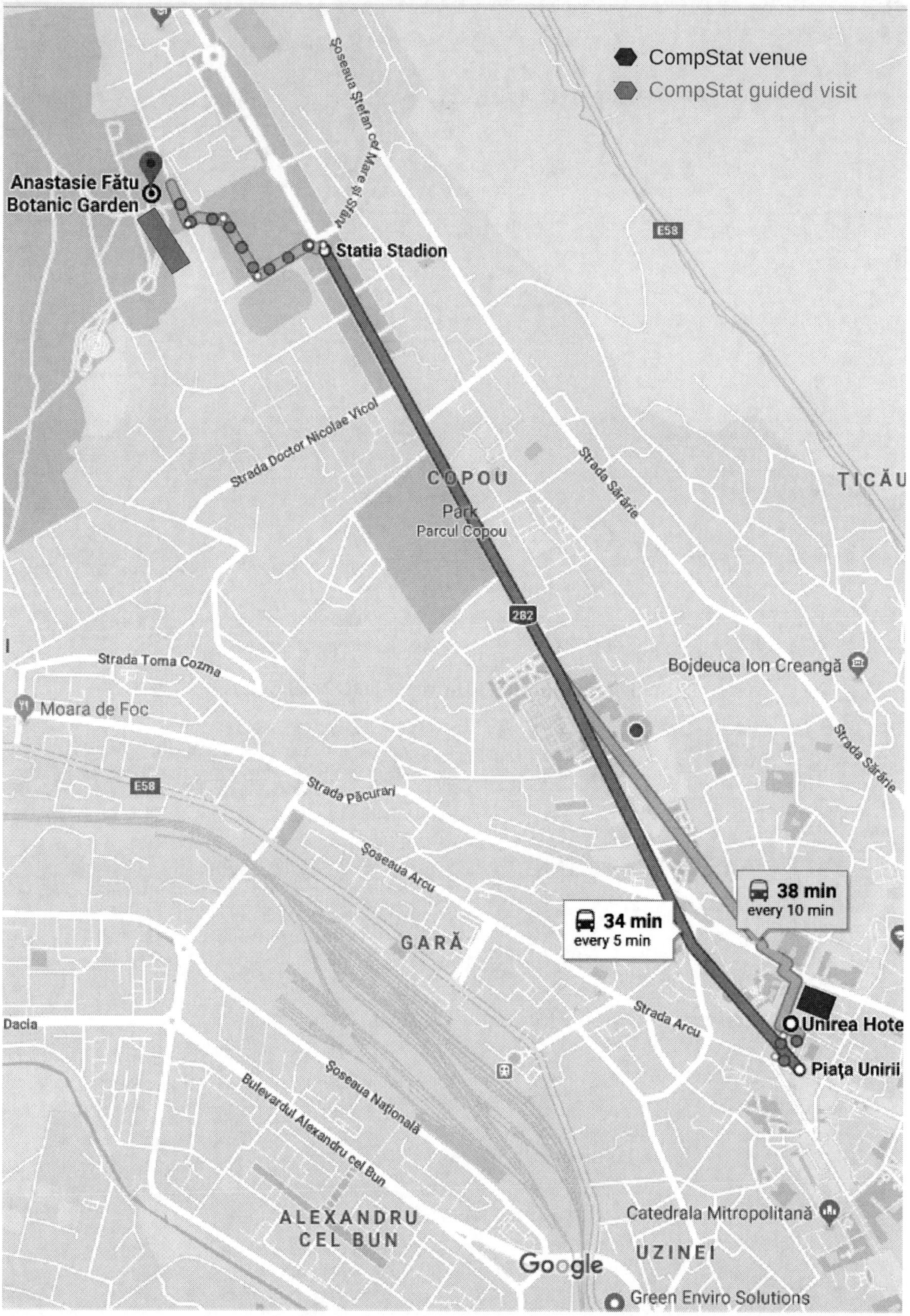

Unirea Hotel - Ground Floor

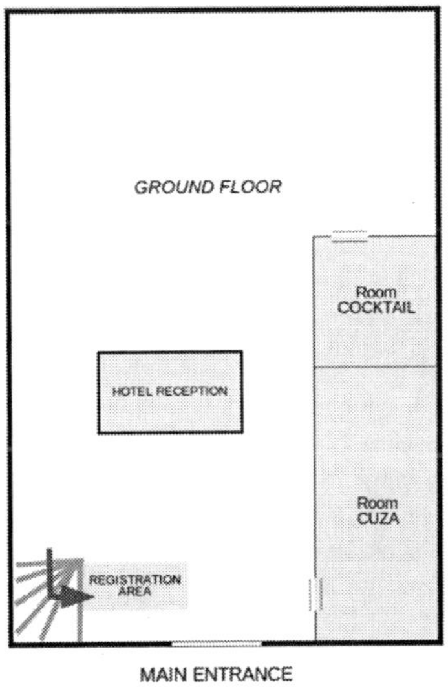

Unirea Hotel - First Floor

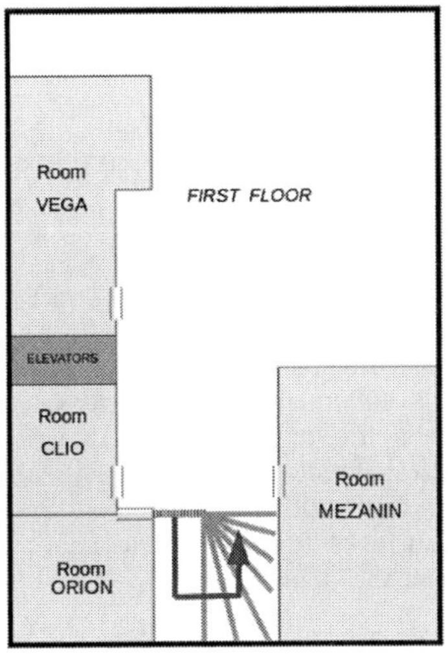

Faculty of Computer Science - Second Floor

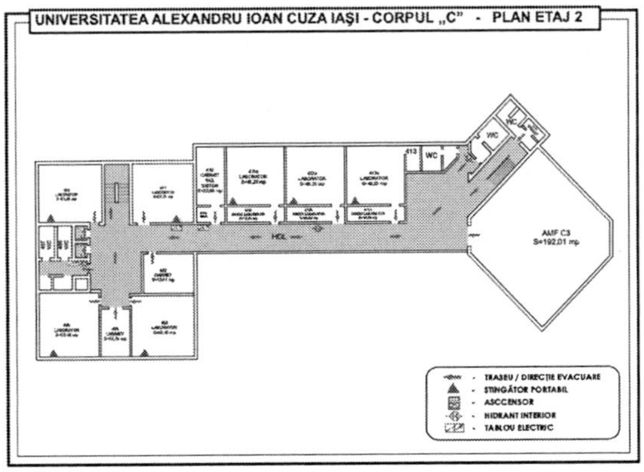

Contents

Tuesday 28.08.2018	09:40 - 10:30	Room: Cuza Center	Chair: Stefan Van Aelst	Keynote talk 1

Fast robust correlation for high dimensional data

Speaker: **Peter Rousseeuw, KU Leuven, Belgium**

The product moment covariance is a cornerstone of multivariate data analysis, from which one can derive correlations, principal components, Mahalanobis distances and many other results. Unfortunately the product moment covariance and the corresponding Pearson correlation are very susceptible to outliers (anomalies) in the data. Several robust measures of covariance have been developed, but few are suitable for the ultrahigh dimensional data that are becoming more prevalent nowadays. For that one needs methods whose computation scales well with the dimension, are guaranteed to yield a positive semidefinite covariance matrix, and are sufficiently robust to outliers as well as sufficiently accurate in the statistical sense of low variability. We construct such methods using data transformation. The resulting approach is simple, fast and widely applicable. We study its robustness by deriving influence functions and breakdown values, and computing the mean squared error on contaminated data. Using these results we select a method that performs well overall, which we call wrapping. It is available in the R package *cellWise*. Wrapping allows a very substantial speedup of the *DetectDeviatingCells* technique for flagging cellwise outliers, which is applied to genomic data with 12,000 variables. Wrapping is able to deal with even higher dimensional data, which is illustrated on color video data with 920,000 dimensions.

Thursday 30.08.2018	11:30 - 12:20	Room: Cuza Center	Chair: Cristian Gatu	Keynote talk 2

Recent developments in algorithmic optimal design of experiments

Speaker: **Steven Gilmour, KCL, United Kingdom**

Optimal design of experiments are implemented in two ways: either a continuous optimal design is found using continuous optimization methods, in which only the proportions of runs made at each support point are found and the continuous design is then rounded to obtain a design for the experiment, or an exact near-optimal design is found using discrete optimization methods, which can only be guaranteed to converge to local optimum, but give a design for the experiment directly. Continuous designs are usually more useful when the number of runs in the experiment is much greater than the number of parameters in the model, especially for nonlinear models. Exact designs are usually more useful for experiments with multiple factors, in which the number of runs is typically not much greater than the number of parameters, especially in linear models. There has been increasing interest in nonlinear models with multiple factors, which are often hybrids of mechanistic and empirical models. In such cases continuous designs can be unsatisfactory, without the rounding becoming a new search in itself, and exact design methods suffer from the need to specify a discrete set of candidate points. New exact design algorithms will be discussed which try to get round these problems. They include using algebraic expressions for the optimality criterion, using the continuous design to suggest candidate points and using continuous multidimensional optimization to avoid the need for candidate points.

Friday 31.08.2018	12:10 - 13:00	Room: Cuza Center	Chair: Erricos John Kontoghiorghes	Keynote talk 3

Generalised autocovariances and spectral estimators

Speaker: **Alessandra Luati, University of Bologna, Italy**

A class of models for the time-varying spectrum of a locally stationary process is introduced. The models are specified in the frequency domain and the class depends on a power parameter that applies to the spectrum so that it can be locally represented by a finite Fourier polynomial. The coefficients of the polynomial are dynamic generalised cepstral coefficients that have an interpretation as generalised autocovariances. The dynamics of the generalised cepstral coefficients are determined according to a linear combination of logistic transition functions of the time index. Estimation is carried out in the frequency domain based on the generalised Whittle likelihood.

Tuesday 28.08.2018 11:00 - 12:30 Parallel Session B – COMPSTAT2018

CI002 Room Cuza Hall COMPOSITIONAL DATA ANALYSIS Chair: Karel Hron

C0251: **The log-ratio methodology: Major concepts, robustness, and practical use**
Presenter: **Peter Filzmoser**, Vienna University of Technology, Austria

In compositional data analysis, the interest is in analyzing relative information rather than the measured or observed values directly. This can be done by considering the log-ratios between all pairs of variables, and - to avoid over-parametrization - by constructing an orthonormal basis describing this information. Since (log-)ratios are taken, one could multiply the values of one observation by a positive constant without changing this relative information. This implies that compositional data analysis is not only limited to data sets with a constant sum of the observations - a frequent misunderstanding. We will introduce some of the major concepts of the log-ratio methodology for compositional data, and present their use in methods for multivariate data analysis. Special attention is given to robust statistical methods.

C0252: **Compositional tables and their coordinate representations**
Presenter: **Karel Hron**, Palacky University, Czech Republic
Co-authors: Kamila Facevicova, Julie Rendlova

A data table which is arranged according to two factors can be considered as a compositional table if the relative structure of the relationship between factors is of primary interest. An example is the total amount of unemployed people, split according to gender and age classes. Analyzed as a compositional table, the relevant information would consist of ratios between different cells of such a table. This perspective is particularly useful when analyzing several compositional tables jointly, where the absolute scale of observations may differ substantially, e.g. if unemployment data are considered from different countries. Within the framework of the logratio methodology, compositional tables can be decomposed into independent and interactive parts, and real orthonormal coordinates that enable statistical processing using standard multivariate methods can be assigned to these parts. The aim is to review recent developments on compositional tables, particularly to construct such orthonormal coordinates that make it possible to decompose the table. As an alternative, coefficients with respect to a generating system (centered logratio coefficients), that are frequently preferred in the logratio methodology due to easy interpretability, will be discussed. Finally, also an outlook to the multi-factorial case will be outlined.

C0298: **A Bayes space approach to the analysis of probability density functions**
Presenter: **Alessandra Menafoglio**, Politecnico di Milano, Italy

In several studies elementary data are aggregated, and then represented through probability density functions (PDFs). For instance, in socio-economic contexts, the age of the population is often described through its distribution (i.e., a population pyramid), whereas, in environmental studies, the soil granularity is typically represented through a particle-size distribution. In all these cases the dataset consists of PDFs, whose proper statistical treatment is key to describe, model and predict the phenomenon under study. Statistical methods for the analysis of PDFs need to account for the infinite-dimensionality of the data, and their inherent constraints. We will discuss the Bayes space viewpoint to the analysis of PDFs, which combine the approaches of functional data analysis and compositional data analysis, through the foundational role of the generalized Aitchison geometry. In this framework, methods for dimensionality reduction, modeling and prediction will be illustrated, with application to studies of industrial and environmental interest.

CO100 Room Mezzanine Lounge RECENT ADVANCES IN BIOSTATISTICS AND BIOCOMPUTING Chair: Yisheng Li

C0170: **An alternative sensitive analysis approach for missing not at random**
Presenter: **Chiu-Hsieh Hsu**, University of Arizona, United States
Co-authors: Chengcheng Hu, Yulei He

Missing mechanism is unverifiable. Often researchers perform sensitivity analysis to evaluate the impact of various missing mechanisms. All the existing sensitivity analysis approaches for missing not at random (MNAR) require to fully specify the relationship between the missing value and the missing probability. The relationship is specified using a selection model, a pattern-mixture model or a shared parameter model. We propose an alternative sensitive analysis approach for MNAR using a nonparametric multiple imputation approach. The proposed approach only requires to specify the correlation between the missing value and the missing probability. The correlation is a standardized measured and can be directly used to indicate the magnitude of MNAR. Numerical results indicate the proposed approach performs well and can be used as an alternative approach for MNAR.

C0305: **A Bayesian dose-finding design in oncology using pharmacokinetic/pharmacodynamic modeling**
Presenter: **Yisheng Li**, The University of Texas MD Anderson Cancer Center, United States
Co-authors: Xiao Su, Peter Mueller, Kim-Anh Do

While a number of phase I dose-finding designs in oncology exist, the commonly used ones are either algorithmic or empirical model based. We propose a new framework for modeling the dose-response relationship via dynamic PK/PD modeling and modeling of the relationship between the pharmacologic effect and a binary toxicity outcome. Inference is implemented in one joint model that encompasses PK, PD and clinical outcome. This modeling framework naturally incorporates the information on dose, schedule and method of administration (e.g., drug formulation and route of administration) in their relationship with toxicity. Simulations show that the performance of the proposed DISCO design on average improves upon those of currently used designs, including the CRM, BOIN and mTPI designs, and a hypothetically optimal non-parametric design in some scenarios. A sensitivity analysis suggests that the performance of the DISCO design is robust with respect to assumptions related to the interindividual variability in the PK. The DISCO design is less expensive and more ethical than existing designs since it makes efficient use of the information from the enrolled patients, and it does not require PK data from all patients or real-time PK analysis. We illustrate the proposed design by applying it to the setting of a phase I trial of a gamma-secretase inhibitor in metastatic or locally advanced solid tumors.

C0223: **Joint scale change models for recurrent events and failure time**
Presenter: **Chiung-Yu Huang**, University of California, San Francisco, United States
Co-authors: Gongjun Xu, Sy Han Chiou, Chiung-Yu Huang, Jun Yan

Recurrent event data arise frequently in various fields such as biomedical sciences, public health, engineering, and social sciences. In many instances, the observation of the recurrent event process can be stopped by the occurrence of a correlated failure event, such as treatment failure and death. We propose a joint scale-change model for the recurrent event process and the failure time, where a shared frailty variable is used to model the association between the two types of outcomes. In contrast to the popular Cox-type joint modeling approaches, the regression parameters in the proposed joint scale-change model have marginal interpretations. The proposed approach is robust in the sense that no parametric assumption is imposed on the distribution of the unobserved frailty and that we do not need the strong Poisson-type assumption for the recurrent event process. We establish consistency and asymptotic normality of the proposed semiparametric estimators under suitable regularity conditions. To estimate the corresponding variances of the estimators, we develop a computationally efficient resampling-based procedure. Simulation studies and an analysis illustrate the performance of the proposed method.

C0158: **Accelerated methods for maximum likelihood estimation in mixed effects models**
Presenter: **Belhal Karimi**, INRIA Saclay - Ecole Polytechnique, France

Co-authors: Marc Lavielle, Eric Moulines

Several improvements of existing methods for maximum likelihood estimation (MLE) in models with latent variables are presented. We focus on mixed effects models where the random effects are latent. In the context of nonlinear mixed effects models, the Stochastic Approximation of the EM algorithm (SAEM) is very efficient and widely used for MLE. We propose an incremental version of the SAEM that accelerates its convergence. Incremental methods have been vastly studied in the context of gradient descent type algorithms where considering a batch of points allows using bigger stepsizes and thus achieving faster convergence. We propose its extension to the SAEM. We consider an MCMC procedure for sampling the random effects and/or estimating their conditional distribution. The choice of the proposal distribution is critical mainly for multidimensional space. New techniques such as SDE-based or Hamiltonian dynamics may be efficient but are difficult to tune and are costly. We propose the use of a multidimensional Gaussian proposal that takes into account the covariance structure of the random effects we want to infer and does not require any tuning. Numerical experiments based on simulated and real data highlight the very good performances of the proposed methods.

CO018 Room Orion Hall SURVEY SAMPLING — Chair: Alina Matei

C0227: **Estimating a counterfactual wage heavy-tailed distribution using survey data**

Presenter: **Alina Matei**, University of Neuchatel, Switzerland

Co-authors: Mihaela Catalina Anastasiade, Yves Tille

The focus is on the framework of the gender wage modelisation using survey data. The wage of an employee is hypothetically a reflection of their characteristics, such as the education level or the previous work experience. It is possible that a man and a woman with the same characteristics get different salaries. To measure the difference in the gender wages the concept of counterfactual distribution is used. A counterfactual distribution is constructed by reweighting the women wage distribution. We provide two parametric methods to estimate the gender wage quantiles and counterfactual wage quantiles, respectively, and estimate their differences. The goal is to capture the shape of the wage distributions and to go beyond the simple mean differences, by determining the estimator of gender wage discrimination at different quantiles. Since, in general, wage distributions are heavy-tailed, the main interest is to model wages by using heavy-tailed distributions like the GB2 distribution. We illustrate the two proposed methods using the GB2 distribution and compare them with other approaches found in the topic-related literature.

C0275: **Bootstrap variance estimation for multistage sampling designs**

Presenter: **Guillaume Chauvet**, ENSAI-IRMAR, France

Multistage sampling designs are commonly used for household surveys. If we wish to perform longitudinal estimations, individuals from the initial sample are followed over time. If we also wish to perform cross-sectional estimations at several times, additional samples are selected at further waves and mixed with the individuals originally selected. Even in the simplest case when estimations are produced at the first time with a single sample, variance estimation is challenging since the different sources of randomness need to be accounted for, along with the needed statistical treatments (correction of unit non-response at the household and at the individual level, correction of item non-response, calibration). We consider a bootstrap solution which accounts for the features of the sampling and estimation process. This bootstrap solution is usually conservative for the true variance, in that the sampling variance tends to be overestimated. The proposed bootstrap is illustrated with examples, and the results of a simulation study are presented.

C0217: **Fast implementation and generalization of Fuller's unequal probability sampling method**

Presenter: **Yves Tille**, University of Neuchatel, Switzerland

Wayne Fuller proposed in 1970 a very ingenious method of sampling with unequal inclusion probabilities. Firstly, doing justice to this precursor paper of Fuller, a very simple and fast implementation of this method is proposed. Secondly, the method is generalized in order to enable the use of a tuning parameter of spreading.

C0231: **Spatial small area smoothing models for handling survey data with nonresponse**

Presenter: **Kevin Watjou**, University Hasselt, Belgium

Co-authors: Christel Faes, Andrew Lawson, Rachel Carroll, Mehreteab Aregay, Yannick Vandendijck, Russell Kirby

Spatial smoothing models play an important role in the field of small area estimation. In the context of complex survey designs, the use of design weights is indispensable in the estimation process. Recently, efforts have been made in the development of spatial smoothing models, in order to obtain reliable estimates of the spatial trend. However, the concept of missing data remains a prevalent problem in the context of spatial trend estimation as estimates are potentially subject to bias. We focus on spatial health surveys where the available information consists of a binary response and its associated design weight. Furthermore, we investigate the impact of nonresponse as missing data on a range of spatial models for different missingness mechanisms and different degrees of nonresponse by means of an extensive simulation study. The computations were done in R, using INLA and other existing packages. The results show that weight adjustment to correct for nonresponse has a beneficial effect on the bias in the missing at random (MAR) setting for all models. Furthermore, we estimate the geographical distribution of perceived health at the district level, based on the 2011 Belgian Health Interview Survey.

CO044 Room Vega Hall BDSPORTS: SPORTS ANALYTICS — Chair: Christophe Ley

C0329: **Modelling the dynamic pattern of surface area in basketball and its effects on team performance**

Presenter: **Rodolfo Metulini**, 1983, Italy

Due to the advent of GPS techniques, a wide range of scientific literature on Sport Science is nowadays devoted to the analysis of players movement in relation to team performance in the context of big data analytics. A specific research question regards whether certain patterns of space among players affect team performance, from both an offensive and a defensive perspective. Using a time series of basketball players coordinates, we focus on the dynamics of the surface area of the five players on the court with a twofold purpose: (i) to give tools allowing a detailed description and analysis of a game with respect to surface areas dynamics and (ii) to investigate its influence on the points made by both the team and the opponent. We propose a three-step procedure integrating different statistical modelling approaches. Specifically, we first employ a Markov Switching Model (MSM) to detect structural changes in the surface area. Then, we perform descriptive analyses in order to highlight associations between regimes and relevant game variables. Finally, we assess the relation between the regime probabilities and the scored points by means of Vector Auto Regressive (VAR)models. We carry out the proposed procedure using real data and, in the analyzed case studies, we find that structural changes are strongly associated to offensive and defensive game phases and that there is some association between the surface area dynamics and the points scored by the team and the opponent.

C0234: **Ranking soccer teams on their current strength: A comparison of maximum likelihood approaches**

Presenter: **Hans Van Eetvelde**, Ghent University, Belgium

Co-authors: Christophe Ley

Different strength-based statistical models are presented which can be used to model soccer match outcomes. The models are of three main types: Bradley-Terry, Independent Poisson and Bivariate Poisson, and their common aspect is that the parameters are estimated via weighted maximum likelihood, the weights being a match importance factor and a time depreciation factor giving less weight to matches that are played a long time ago. Their predictive performance is compared via the Rank Probability Score and the log loss. We will consider two cases: the Premier league as an example of a domestic league and the case of national teams.

C0314: **Passing networks and game style in football teams: Evidences from European champions league**
Presenter: **Lucio Palazzo**, University of Salerno, Italy
Co-authors: Riccardo Ievoli, Giarncarlo Ragozini

Summary statistics of football matches (such as final score, possession and percentage of completed passes) give in general poor information about style of play seen on the pitch. On such bases, it is generally difficult to quantify how teams are different from each other. The focus is on the analysis of weighted and directed passing network of football teams. In particular, descriptive measures and structural features of networks are analyzed in order to evaluate different team strategies in terms of team passing behaviour by using passage interactions among players. The main contribution is twofold: on one side is showed how structural properties measured through triadic census are able to distinguish among different styles of play. Useful graphic visualization for the comparison of teams and their own level of interaction between players are provided. On the other hand, passing network indices and structural properties are used to better predict probability of winning the match. Data include team passing network regarding 96 matches in the group stage of UEFA Champions League 2016-2017, involving 32 European teams.

C0397: **Model-based profiling of sport preferences**
Presenter: **Rosaria Simone**, University of Naples Federico II, Italy

Sport preferences and related attitudes can be analysed by collecting ranking or rating data. In these circumstances, it is advisable to adopt a flexible modelling of such discrete distributions that enables the understanding of the evaluation process and the derivation of response profiles. This issue is crucial for marketing purposes, for instance, or when social and behavioural policies have to be addressed as in case sport participation and engagement are investigated. Then, suitable mixture distributions can be specified either to determine clusters of opposite evaluations (for instance, a mixture of inverse hypergeometric models), or to identify structured and guessing response patterns, by choosing a framework in which model parameters are directly linked to explanatory variables. Under this perspective, CUB models lend themselves to advantageous interpretation of results by explicitly accounting for uncertainty. This paradigm can be successfully adapted to incorporate more involved response schemes and to design model-based regression trees and disentangle explanatory features of variables at different subsetting levels. A comparative overview of alternative models based on flexible mixtures of discrete distributions is presented on survey data collected within the BDsport project of the BoDai Lab of University of Brescia, Italy.

Tuesday 28.08.2018 14:15 - 15:45 Parallel Session C – COMPSTAT2018

CI008 Room Cuza Hall COMPUTATIONAL DEPENDENCE MODELING AND CHANGE-POINT DETECTION Chair: Ivan Kojadinovic

C0160: **Visualization of dependence in high-dimensional data with R**
Presenter: **Marius Hofert**, University of Waterloo, Canada
The focus is on Data Visualization as part of Data Science. The question of how high-dimensional data can be visualized is raised. The notion of a zenpath and a zenplot is introduced to search and visualize high-dimensional data for model building and statistical inference. By using any measure of "interestingness", a zenpath can construct paths through pairs of variables in different ways, which can then be laid out and displayed by a zenplot. Zenpaths and zenplots are useful tools for exploring dependence in high-dimensional data, for example, from the realm of finance, insurance and quantitative risk management. All presented algorithms are implemented using the R package zenplots.

C0258: **Robust change point tests using bounded transformations**
Presenter: **Alexander Duerre**, TU Dortmund, Germany
Co-authors: Roland Fried, Daniel Vogel
Classical moment based change point tests like the cusum test are very powerful under Gaussian time series with no more than one change point but behave poorly under heavy tailed distributions and corrupted data. A new class of robust change point tests based on cusum statistics of robustly transformed observations is proposed. This framework is very flexible, depending on the used transformation one can detect amongst others changes in the mean, scale or dependence of a possibly multivariate time series. The calculation of p-values can be simplified by using asymptotics which yields a computational complexity of $T\log(T)$ where T is the number of observations. Symmetrization leads to change point tests based on U-Statistics, which are more powerful under Gaussianity but less robust and have a complexity of T^2. The application of both approaches is illustrated on the basis of some examples which are evaluated using the statistical software R.

C0259: **Testing for changes in the distribution function, cross-sectional and serial dependence of multivariate time series**
Presenter: **Ivan Kojadinovic**, CNRS UMR 5142 LMA University of Pau, France
Cumulative sum (CUSUM) change-point tests based on U-statistics, empirical distribution functions, cross-sectional and serial empirical copulas are presented. After a brief exposition of the underlying asymptotic and resampling theory, illustrations based on the R package npcp are presented and implementation aspects (such as the computation of a key bandwidth parameter) are discussed.

CO109 Room Mezzanine Lounge ADVANCES IN INDUSTRIAL STATISTICS Chair: Chun-houh Chen

C0179: **Mis-specification analysis of a pH acceleration model**
Presenter: **S-T Tseng**, National Tsing-Hua University, Taiwan
Shelf-life prediction of highly reliable nano-sol products is an interesting research topic. Recently, a pH acceleration model has been proposed in the literature for the time evolution of nano-particle distributions. There are two approaches for modelling the particle-size distribution, by using either a parametric approach (mixture-normal distribution) or a non-parametric (distribution free) approach. Their shelf-life predictions, however, have significant differences. Therefore, quantifying the effects of model mis-specification of particle size distribution on the shelf-life prediction of nano-sol products turned out to be a challenging issue. The accuracy and precision of the shelf-life prediction were obtained analytically if the true particle-size follows a mixture-normal distribution, but wrongly treated as a distribution free model. For the given set of data, the shelf-life prediction may be under-estimated up to 13.66%, while its relative variation may be inflated up to 13.35 times by using a non-parametric approach. In addition, simulations show that when the sample size and the measurement times are small, these effects are moderately large and cannot be neglected.

C0192: **Optimal sample size allocation for accelerated life tests of a series system**
Presenter: **Tsai-Hung Fan**, National Central University, Taiwan
In accelerated life tests of system reliability, the sample size allocation under different stress levels could affect the accuracy of the reliability inference. Given three stress levels of an accelerated variable, the issue on the optimal allocation of an accelerated life test of series systems is tackled. It turns out that the objective functions frequently are of the form of the product of second elementary symmetric functions. We fist derive the sufficient condition when the optimal plan is reduced to a two-level test with equal sample size allocated at the lowest and the highest levels for systems connected by two components. Under independent exponential life time distributions of the components, more specific results, such as the relative efficiency of the three-level uniform design to the optimal allocation, are developed. The results are also demonstrated and justified by a real example. Generalization to a multi-component series system is conjectured and verified by numerical results.

C0228: **A non-parametric approach for monitoring linear profiles using spatial rank-based regression**
Presenter: **Longcheen Huwang**, National Tsing Hua University, Taiwan
Profile monitoring has been recently considered as one of the most promising areas of research in statistical process monitoring (SPM). The monitoring of linear profiles is the most popular one because the relationship between the dependent variable and the independent variables is easy to describe by linearity, in addition to its flexibility and simplicity. Furthermore, almost all existing charting schemes for monitoring linear profiles assume that error terms are normally distributed. In some applications, however, the normality assumption of error terms is not justified. This makes the existing charting schemes not only inappropriate, but also less efficient for monitoring linear profiles. We propose a non-parametric charting method for monitoring linear profiles where the error terms are not normally distributed based on the spatial rank-based regression. The charting scheme applies the exponentially weighted moving average (EWMA) to the spatial rank of the vector of the Wilcoxon-type rank-based estimators of regression coefficients and a transformed error variance estimator. Performance properties of the proposed charting scheme are evaluated and compared with an existing non-parametric charting method in terms of the in-control (IC) and out-of-control (OOC) average run length (ARL). Finally, a real example is used to demonstrate the applicability and implementation of the proposed charting scheme.

C0273: **Optimal doubling burn-in policy based on tweedie processes with applications to degradation data**
Presenter: **Chien-Yu Peng**, Academia Sinica, Taiwan
In the current competitive marketplace, manufacturers need to screen weak products in a short period of time. It is a challenge for manufacturers to implement a burn-in test that can screen out the weak products quickly and efficiently. When selecting an approach to determine the duration of the burn-in, one could build a criterion aiming to minimize the burn-in cost. In practice, when the optimal burn-in time is unreasonable (e.g., time 0) due to minimizing the cost, this means that the burn-in procedure is unnecessary to perform for manufacturers. We propose an optimal doubling burn-in policy to improve the predicament without additional experiments. The advantage of the proposed policy is to simultaneously determine the optimal burn-in time and the optimal cutoff point for classifying weak and strong components from the production. In addition, a new degradation model based on a Tweedie mixture process is used for a burn-in test. The proposed burn-in procedure is applied to a real data.

C0456: **Association of cardiovascular responses in mice with PM2.5 air pollution from traffic and industry sources**
Presenter: **Jing-Shiang Hwang**, Academia Sinica, Taiwan
For concerns about the health of athletes and international visitors to 2008 Olympic Games in Beijing, the government mitigated the ambient air pollution by relocating, limiting or temporarily closing highly polluting, energy-intensive facilities in and around the city, and reducing vehicle

usage by elaborate traffic regulations. These air quality interventions provided us a unique opportunity to assess the effect of reduction in fine particles on cardiovascular responses. We propose Bayesian analysis with models including weather parameters to identify fine particulate matter (PM2.5) sources and estimate their contributions to the ambient air pollution in Beijing. The estimated contributions were brought into mixed-effects models as exposures for examining the association of cardiovascular responses of the exposed mice in an experiment of four months. We found that while overall PM2.5 mass has a negative effect onto heart rate, interestingly, traffic related and mixed industrial categories increased the heart rate on the day of exposure but had significant negative contribution on day lag 1 and 2; while oil combustion and secondary sulphate had positive effect on day lag 2.

CO064 Room Vega Hall ROBUST STATISTICS Chair: Anthony Atkinson

C0190: A computationally feasible algorithm for robust clustering under determinant-and-shape constraints

Presenter: **Luis Angel Garcia-Escudero**, Universidad de Valladolid, Spain

Co-authors: Andrea Cerioli, Agustin Mayo-Iscar, Marco Riani

The most widely applied approaches in model-based clustering are based on the maximization of classification and mixture likelihoods. Under standard normal assumptions, these likelihood maximizations are mathematically ill-posed problems without appropriate constraints on the components' scatter matrices. Moreover, non-interesting or "spurious" solutions are often detected by traditional CEM and EM algorithms designed for them. This is also the case when robustifying them through the use of trimmed likelihoods. An upper bound on the ratio between the largest and smallest determinants for the components' scatter matrices is apparently a sensible way to overcome those degeneracy troubles. Unfortunately, this type of constraints, although affine equivariant, does not always avoid spurious solutions and, consequently, robustness cannot be guaranteed. On the other hand, we will see how some additionally added constraints on the components shape elements actually serve to cope with those degeneracy issues. The combination of trimming and this new type of constraints results in an (almost) affine equivariant robust model-based clustering approach. A computationally feasible algorithm is proposed for this new approach.

C0168: Robustifying mean regression

Presenter: **Qiang Sun**, University of Toronto, United States

Big data can easily be contaminated by outliers and heavy-tailed errors, which makes many conventional methods inadequate. To address this challenge, we propose the adaptive Huber regression for robust estimation and inference. The key observation is that the robustification parameter should adapt to the sample size, dimension and moments for optimal tradeoff between bias and robustness. Our framework is able to handle heavy-tailed errors with low bounded moments. We establish a sharp phase transition for robust estimation of regression parameters in both low and high dimensions with respect to the order of moments, and the transition is smooth and optimal. In addition, we extend the methodology to allow both heavy-tailed predictors and error variables. Simulation studies and an application to cancer cell lines lend further support to our obtained theory. Extension to other models will also be discussed.

C0287: The use of prior information in very robust regression for the monitoring of EU sensitive products

Presenter: **Aldo Corbellini**, Faculty of Economics - University of Parma, Italy

Co-authors: Andrea Cerasa, Francesca Torti

It is well known in the literature that overlooking outliers might have a severe impact on the analysis, leading to biased and inconsistent estimates and misleading inference. We develop a method of detecting the patterns of outliers that indicate systematic deviations in pricing. Since the data only become available year by year, we develop a combination of very robust regression and the use of 'cleaned' prior information from earlier years, which leads to sharp indication of market price anomalies. As a method of very robust regression, we use the Forward Search. A form of empirical Bayesian analysis is extended to incorporate different amounts of prior information about the parameters of the linear regression model and the error variance. As an example we use yearly imports and exports of goods traded by the European Union. We provide a solution to the resulting big data problem, which requires analysis with the minimum of human intervention.

C0289: Monitoring robust estimates for compositional data

Presenter: **Valentin Todorov**, UNIDO, Austria

The monitoring of robust estimates computed over a range of key parameter values is a recent technique. Through this approach the diagnostic tools of choice can be tuned in such a way that highly robust estimators which are as efficient as possible are obtained. Key tool for detection of multivariate outliers and for monitoring of robust estimates are the scaled Mahalanobis distances and statistics related to these distances. However, the results obtained with this tool in case of compositional data might be unrealistic. Compositional data are closed data, i.e. they sum up to a constant value (1 or 100 per cent or any other constant). This constraint makes it necessary to find a transformation of the data from the so called simplex sample space to the usual real space. To illustrate the problem of monitoring compositional data we start with a simple example and then analyze a real life data set presenting the technological structure of manufactured exports which, as an indicator of their quality, is an important criterion for understanding the relative position of countries measured by their industrial competitiveness. The analysis is conducted with the R package fsdaR, which makes the analytical and graphical tools already available in the MATLAB FSDA library available for R users.

CG116 Room Cocktail Hall CLUSTERING AND CLASSIFICATION Chair: Christophe Biernacki

C0363: Classification trees with Gini samples splits

Presenter: **Amirah Alharthi**, University of Leeds, United Kingdom

Many numerical studies indicate that bagged decision stumps preform more accurately than a single stump. We will introduce a new stump-based ensemble method for classification which is: A forest of stumps 'Gini-Sampled Splits'. A stump within this forest uses a split that is generated from transformed Gini indices for each possible cut points. The choice of variable which is chosen on which to generate a split has a probability proportional to that variable Gini index values. The final decision of these stumps is aggregated using weighted vote rather than majority vote. We compared between this method and other tree-based ensemble classification methods in terms of the accuracy and the results are promising.

C0409: Classification based on dissimilarities towards prototypes

Presenter: **Beibei Yuan**, Leiden University, Netherlands

Co-authors: Willem Heiser, Mark De Rooij

The delta-machine, a statistical learning tool for classification based on dissimilarities or distances, is introduced. In the first step dissimilarities between profiles of the objects and a set of selected exemplars or prototypes in the predictor space are computed. In the second step, these dissimilarities take the role as predictors in a logistic regression to build classification rules. This procedure leads to nonlinear classification boundaries in the original predictor space. We discuss the delta-machine with mixed nominal, ordinal, and numerical predictor variables. Two dissimilarity measures are distinguished: the Euclidean distance and the Gower measure. The first is a general distance measure, while the second is a tailored dissimilarity measure for mixed type of variables. Using simulation studies we compared the performance of the two dissimilarity measures in the delta-machine using three types of artificial data. The simulation studies showed that overall the Euclidean distance and the Gower measure had similar performances in terms of the accuracy, but in some conditions the Euclidean distance outperformed the Gower measure. Furthermore, we will show the classification performance of the delta-machine in comparison to three other classification methods on an empirical example.

C0344: **Active learning classification with variable selection**
Presenter: **Yuan-chin Chang**, Academia Sinica, Taiwan

Active learning usually refers to certain kinds of learning methods that could select learning subjects sequentially. We will discuss some logistic model-based active learning methods with variable selection features in addition to subject selection strategies. We adopt a batch subject selection strategy with a modified sequential experimental design method and simultaneously conduct a greedy variable selection procedure such that we can update the classification model with all labeled training subjects. Another method is based on the stochastic regression where subjects are selected one-by-one adaptively and variables are identified sequentially. We repeat these algorithms repeatedly until a corresponding stopping criterion is reached. Our numerical results confirm that the proposed procedures can produce competitive performances with a smaller training size and a more compact model compared with that of the classifier trained with all variables and a full data set.

C0421: **Automatic detection of clusters by lifting**
Presenter: **Nebahat Bozkus**, University of Leeds, United Kingdom
Co-authors: Stuart Barber

In clustering, the aim is to group related objects together and an important question is how to detect the number of groups in a data set. Many methods have been proposed which capture the number of groups quite well if the groups are well separated and regularly shaped. However, if groups overlap or have unusual shapes, the performance of these methods deteriorates. We propose a new method based on a multiscale technique called lifting which has recently been developed to extend the 'denoising' abilities of wavelets to data on irregular structures. The method seeks for the best representation of the clustering pattern by checking all possible clustering schemes in a tree. After denoising the tree, if the leaves under a node are all close enough to their centroid for the deviations to be explained as 'noise', we label those leaves as forming a cluster. The proposed method automatically decides how much departure can be allowed from the centroid of each cluster. The behaviour of the method will be illustrated using some phylogenetic data sets.

CG043 Room Orion Hall ADVANCES IN BAYESIAN STATISTICS AND MARKOV CHAIN MODELS Chair: Miguel de Carvalho

C0431: **Bayesian sequential marginalization of a state space model for estimating motor unit numbers**
Presenter: **Gareth Ridall**, Lancaster University, United Kingdom

This application comes from the field of neurology where an estimate the number of units supplying a muscle group is required. An increasing stimulus applied at the nerve each motor unit of the axon bundle is activated with increasing probability and the cumulative muscular response is recorded. A state space model with increasing dimension is used to model the increasing muscular response with stimulus. The observations are assumed Gaussian, conditional on the parameters and the latent binary firing indicators. In our state space model sufficient statistics and approximate sufficient statistics are used to model the state and measurement processes respectively. An efficient proposal system is used for the current firing indicators. These are re-weighted sequentially using the evidence as observations arrive. Lastly a residual resampling step is used to keep the numbers of possible weighted histories to a manageable number. Our methodology is substantially faster than Reversible jump Markov Chain Monte Carlo and is able to account for drift in the parameters for the measurement process.

C0413: **Bayesian estimation of a decreasing density**
Presenter: **Lixue Pang**, Delft University of Technology, Netherlands

Suppose $X_1,\dots,X_n$ is a random sample from a bounded and decreasing density f_0 on $[0,\infty)$. We are interested in estimating f_0, with special interest in $f_0(0)$. This problem is encountered in various statistical applications and has gained quite some attention in the statistical literature. It is well known that the maximum likelihood estimator is inconsistent at zero. This has led several authors to propose alternative estimators which are consistent. As any decreasing density can be represented as a scale mixture of uniform densities, a Bayesian estimator is obtained by endowing the mixture distribution with the Dirichlet process prior. Assuming this prior, we derive contraction rates of the posterior density at zero. We also address computational aspects of the problem and show how draws from the posterior can be obtained using Gibbs sampling. By a simulation study, we compare the behavior of various proposed methods for estimating $f_0(0)$. We further apply the algorithm to current duration data, where we construct pointwise credible regions for the density and distribution functions.

C0327: **A new augmented mixed regression model for proportions**
Presenter: **Agnese Maria Di Brisco**, University of Milano Bicocca, Italy
Co-authors: Sonia Migliorati

A new mixture regression model is proposed to cope with bounded continuous outcomes lying in the closed interval [0,1]. An increasingly common approach to deal with them is the beta regression model which allows for heteroskedasticity and asymmetry. Nevertheless, this model has two main limitations, namely the inability to represent a wide range of phenomena (bimodality, heavy tails and outlying observations), and the failure in modeling values at the boundary of the support. To overcome these limitations, a new regression model is proposed, which is based on a special mixture of two betas (referred to as flexible beta) sharing the same precision parameter but displaying two distinct component means subject to an inequality constraint. This distribution shows strong identifiability and a.s. likelihood boundedness from above, which facilitate its computational tractability. In addition, the distribution is augmented by adding positive probabilities of occurrences of zeros and /or ones. Thus, the final model (augmented flexible beta) is based on a mixed discrete-continuous density, the continuous part of which is itself a mixture. Intensive simulation studies show the good fit of our new regression model in comparison with other models. Inferential issues are dealt with by a (Bayesian) Hamiltonian Monte Carlo algorithm.

C0191: **The weighted Markov chain probability model: Forecasting discrete time sequence data**
Presenter: **Nihan Potas**, Gazi University, Turkey

Statistical theory for the weighted Markov Chain Probability Model of presented. The aim is to discuss the theory and the application of these models to sets of time sequence data which will be summarized using the contingency table form. In the real data application, the grades point average for 8 semesters and cumulative-grades point average for 4 years of 1217 under-graduate students, beginning in the academic year 2013-2014, studying in Faculty of Political Science, Science and Engineering departments of Ankara University were used. Whether the change in students achievement status measurable in8 semesters and 4 years were evaluated. Markov chains can be an effective method for the description of the models, which will enhance capturing the use of forecast dynamic behavior with connection to the stochastic component. Such a forecasting analysis depends on the Markov Chain Theory, which is widely known, as it needs to defect correction that can be used to solve the inaccuracy and impracticability problems likely to arise from the forecast. With this perspective, the weight values for every state can be calculated using the Weighted Markov Chain Transition probability. The analysis clearly illustrates the benefits of weighted Markov Chain Probability model which is validated through accurate and reliable results obtained.

Tuesday 28.08.2018 16:15 - 17:45 Parallel Session D – COMPSTAT2018

CI088 Room Cuza Hall ADVANCES IN HIGH-DIMENSIONAL AND NONPARAMETRIC STATISTICS Chair: Johannes Lederer

C0229: **On the Hodges-Lehmann estimator in a location mixture model**

Presenter: **Fadoua Balabdaoui**, ETH Zurich, Switzerland

The aim is to derive the exact limit distribution of the Hogdes-Lehmann estimator, considered in the semi-parametric model of a location mixture of symmetric distributions. We give sufficient conditions on the true symmetric component for the weak convergence to hold. As already expected, the limit distribution is that of a three-dimensional centered Gaussian distribution. The variance-covariance matrix can be calculated using the known covariance structure of a standard Brownian Bridge. The examples we used to illustrate the theory indicate that the estimator is not to be advocated when the mixture components are not well separated.

C0207: **Fridge: Focused fine-tuning of ridge regression for personalized predictions**

Presenter: **Kristoffer Hellton**, University of Oslo, Norway

Penalized regression methods, depending on one or more tuning parameters, require fine-tuning to achieve optimal prediction performance. For ridge regression, there exist numerous approaches with cross-validation as the standard procedure, but common for all is that one single parameter is chosen for all future predictions. To better adapt to heterogeneity in high-dimensional data, we propose a focused ridge regression, the fridge procedure, with a unique tuning parameter for each covariate vector for which we wish to produce a prediction. The covariate vector specific tuning parameter is defined as the minimizer of the theoretical mean square prediction error, which is explicitly given in case of ridge regression. We propose to estimate the resulting tuning parameter through a plugin approach, and for high-dimensional data, ridge regression with cross-validation is used as the plugin estimate. The procedure is extended to logistic ridge regression by utilizing parametric bootstrap. Simulations show that fridge gives lower average prediction error than standard ridge regression in heterogeneous data, and we illustrate the method in an application of personalized medicine, predicting individual disease risk based on gene expression data.

C0153: **Fresh ideas for tuning parameter calibration**

Presenter: **Johannes Lederer**, University of Washington, United States

Regularization is essential for analyzing the large and complex data that are generated in economics, neuroscience, astronomy, and many other fields. However, Lasso, Ridge Regression, Graphical Lasso, and other regularized methods depend on tuning parameters that are difficult to calibrate both in theory and in practice. We present two approaches to this challenge. The first approach is based on a testing scheme and is to date the only method that is equipped with both fast algorithms and optimal finite sample guarantees. The second approach is based on the minimization of an objective function that avoids tuning parameters altogether. We show that, quite surprisingly, this estimator can be computed efficiently despite it being highly non-convex.

CO111 Room Mezzanine Lounge YS-ISI SESSION: STATISTICAL COMPUTING AND DATA VISUALIZATION Chair: Han-Ming Wu

C0194: **Sensitivity analysis and visualization for functional data**

Presenter: **Yufen Huang**, National Cheng Kung University, Taiwan

The presence of outliers in functional data analysis can greatly influence the results on modeling and forecasting and may lead to the inaccurate conclusion. Hence, detection of such outliers becomes an essential task. Visualization of data not only plays a vital role in discovering the features of data before applying statistical models and summary statistics but also is an auxiliary tool in identifying outliers. The research involved visualization and sensitivity analysis for functional data has not yet received much attention in the literature to date. We propose a method which combines the influence function with the iteration scheme for identifying outliers in functional data, and we develop new visualization tools for displaying features and grasping the outliers in functional data. Comparisons between our proposed methods with the existing methods are also investigated. Finally, we illustrate these proposed methods with simulation studies and real data examples.

C0240: **Decision thresholds using Hilbert-Huang transforms in fMRI applications**

Presenter: **Po-Chih Kuo**, Institute of Statistical Science, Academia Sinica, Taiwan

Co-authors: Michelle Liou, Philip Cheng

Statistical decision on the brain activation maps in functional magnetic resonance imaging (fMRI) time series requires two steps: (1) calculating the statistics in brain regions; (2) thresholding the statistics using a criterion. In Step (2), parametric methods usually make an assumption on the asymptotic distributions of statistics. Alternately, nonparametric procedures determine thresholds using surrogate distributions. For example, a surrogate data approach randomly permutes the phases of raw time series based on Fourier transform which preserves the stationary structure of data. However, spontaneous and induced brain responses in the real world are non-stationary. To incorporate possible non-stationarity in fMRI time series, a randomization procedure based on the Hilbert-Huang transforms is proposed. Two fMRI datasets with either stationary or non-stationary properties were used in our experiment. The significance of individual voxels was determined by comparing the empirical data against the surrogate data distribution. When compared with the results using conventional phase-randomization and wavelet-based permutation methods, the proposed method provided activation maps revealing essential brain regions while filtering out noises in the white matter. Our work shows the importance of considering the non-stationary nature of fMRI time series when selecting resampling methods for probing brain activity or functional networks in real-life fMRI experiments.

C0246: **Visualization and identification of agonistic interaction hepatitis B and C interaction on hepatocellular carcinoma**

Presenter: **Sheng-Hsuan Lin**, Institute of Statistics, Taiwan

Co-authors: Yen-Tsung Huang, Wen-Chung Lee, Hwai-I Yang

Sufficient-component cause (SCC) framework, as one of the most polished techniques for the methodology development of causal inference, has the advantage of visualizing the interaction effect by synergism or antagonism. However, it is well known that statistical interaction occurs even there is no synergism and antagonism, and vice versa. We propose a modified version of SCC, termed exclusive sufficient causal (eSCC) model, and incorporate this model to both counterfactual and DAGs framework. The causal effects can be interpreted as the additive probabilities of conditions under eSCC. When two exposures of interest are considered, eSCC can visualize the existence of agonism, one important subtype of interaction other than synergism and antagonism. We further propose four approaches that suffice to identify and estimate the agonistic interaction by empirical data. We applied the proposed methods to quantify the agonism of Hepatitis B and C viruses (HBV and HCV) infections on liver cancer using a Taiwanese cohort study ($n = 23,820$). The result demonstrates that agonistic interaction is more dominant compared with synergistic interaction, which explains the findings that the dual infected patients do not have a significantly higher risk of liver cancer than those with single infection. This method fills the gap between causal interaction and mechanistic interaction and contributes to a comprehensive understanding of mechanistic investigation.

C0172: **The extension of isometric feature mapping for interval-valued symbolic data**

Presenter: **Han-Ming Wu**, National Taipei University, Taiwan

The dimension reduction of the interval-valued data is one of the active research topics in symbolic data analysis (SDA). The main thread has been focused on the extensions of the linear algorithms such as the principal component analysis (PCA) and the sliced inverse regression (SIR).

We extend the isometric feature mapping (ISOMAP) to the interval-valued data which we called interval ISOMAP (iISOMAP). ISOMAP is a global geometric framework for nonlinear dimensionality reduction (NLDR) techniques using the shortest-path distance in a neighbor graph. The ISOMAP algorithm advances PCA and the multidimensional scaling (MDS) by providing a better understanding of the data's intrinsic structure. Applying interval MDS to the estimation of the geodesic distance between interval data points is the key step of the ISOMAP. For the estimation of the geodesic distance between interval type symbolic objects, we compare the various input distance measures proposed previously. The maximum covering area rectangle (MCAR) method is used to display the interval objects onto a 2D NLDR subspace in order to visualize the geometric structure of a nonlinear manifold dataset. We evaluate the method for the low-dimensional discriminative and visualization purposes by means of the simulation studies and real data sets. The comparison with those obtained with the symbolic PCA and the symbolic MDS were also reported.

CG017 Room Cocktail Hall FUNCTIONAL DATA ANALYSIS — Chair: Stefan Van Aelst

C0388: A multiple functional linear model
Presenter: **Dominique Katshunga**, University of Cape Town, South Africa

The increased need to base statistical models on good quality data has led to the approaches of functional data and functional data analysis (FDA). Functional data refer to data providing information about curves and surfaces varying over a continuum, with smoothness being the key feature. Functional data analysis has enjoyed a considerable amount of attention in the literature under its exploratory and confirmatory statistical analyses approaches. However, under the predictive approach, functional data analysis is still an emerging field. As a part of this emerging field, this discussion focusses on extending the functional linear models (FLM) for functional responses to accommodate more than one covariates. Emphasis is put on fitting the extended models using the backfitting technique. As an illustration, a simulation study is considered.

C0394: Gaussian process methods for nonparametric functional regression with mixed predictors
Presenter: **Bo Wang**, University of Leicester, United Kingdom
Co-authors: Aiping Xu

The aim is to propose Gaussian process methods for nonparametric functional regression for both scalar and functional responses with mixed multidimensional functional and scalar predictors. The proposed models allow the response variables depending on the entire trajectories of the functional predictors. They inherit the desirable properties of Gaussian process regression, and can naturally accommodate both scalar and functional variables as the predictors, as well as easy to obtain and express uncertainty in predictions. The numerical experiments show that the proposed methods significantly outperform the competing models, and their usefulness is also demonstrated by the application to two real datasets.

C0353: Quantifying the closeness to a set of random curves via the mean marginal likelihood
Presenter: **Cedric Rommel**, INRIA, Polytechnique, France
Co-authors: Frederic Bonnans, Pierre Martinon, Baptiste Gregorutti

The problem of quantifying the closeness of a newly observed curve to a given sample of random functions is tackled, when it is supposed that they have been sampled from the same distribution. We define a probabilistic criterion for such a purpose, based on the marginal density functions of an underlying random process. For practical applications, a class of estimators based on the aggregation of multivariate density estimators is introduced and proved to be consistent. We illustrate the effectiveness of our estimators, as well as the practical usefulness of the proposed criterion, by applying our method to a dataset of real aircraft trajectories.

C0395: Model based functional clustering of varved lake sediments
Presenter: **Per Arnqvist**, Umea University, Sweden
Co-authors: Sara Sjostedt de Luna

Climate and environmental changes are today widely discussed, and in particular the impact of human activity. To understand variations in past climate over longer time periods, historical documents, year rings from trees, ice cores from glaciers as well as lake and sea sediments are being used. We introduce a model based functional cluster analysis, giving us possibility to use both the functional form and covariates in our analysis. It also allows us to model the dependency of the chosen basis coefficients and the covariates. We allow for different covariance structure within each cluster and give suggestions on how to determine how many clusters to use. In particular we analyze varved sediment from lake Kassjon (N Sweden) which cover more than 6400 years.

CG114 Room Clio Hall SEMIPARAMETRIC METHODS AND ASYMPTOTICS — Chair: Frank Eriksson

C0400: Targeted causal estimation in continuous time
Presenter: **Helene Charlotte Rytgaard**, University of Copenhagen, Denmark

The aim is to discuss aspects of a continuous time generalization of longitudinal targeted minimum loss-based estimation (TMLE). TMLE is a framework for estimation of causal parameters that combines data-adaptive estimation with a targeting procedure tailored to optimal estimation of a specific low-dimensional parameter of interest. The existing TMLE methods for longitudinal data rely on a discrete data structure where observations are made on the same grid points of time for all subjects. In most realistic settings, however, both exposure and outcome can happen at arbitrary points in time. We propose a unified methodology relying on counting process modeling to handle the data exactly as they are observed. This involves construction of hazard-based initial estimators of the components of the partial likelihood and an extension of the targeting procedure that combines updating steps of intensities and conditional expectations. Potential applications include analysis of large-scale registry databases with regularly updated measurements of all members of a population over large timespans.

C0424: Large sample results for frequentist multiple imputation for Cox regression with missing covariate data
Presenter: **Frank Eriksson**, University of Copenhagen, Denmark
Co-authors: Torben Martinussen, Soren Feodor Nielsen

Incomplete information on explanatory variables is commonly encountered in studies of possibly censored event times. A popular approach to deal with partially observed covariates is multiple imputation, where a number of completed data sets that can be analyzed by standard complete data methods are obtained by imputing missing values from an appropriate distribution. Using a consistent and asymptotically linear but inefficient initial estimator, we impute missing values conditional on the observed data ensuring compatibility with a Cox regression model. We show that estimators of both the finite-dimensional regression parameter and the infinite-dimensional cumulative baseline hazard parameter by Cox regression applied to the completed data sets are consistent and weak convergence is established. We derive a consistent estimator of the covariance operator. Simulation studies and an application to a study on survival after treatment for liver cirrhosis show that the estimators perform well with moderate sample sizes and indicate that iterating the multiple-imputation estimator increases the precision.

C0412: Resampling methods for an adequate tail index estimation
Presenter: **Manuela Neves**, FCiencias.ID, Universidade de Lisboa and CEAUL, Portugal
Co-authors: Ivette Gomes, Helena Penalva, Frederico Caeiro

In Statistics of Extremes the extreme value index (EVI) is a central parameter. We use bootstrapping schemes to perform the choice of two nuisance parameters that appear in a recent class of EVI-estimators. Given a random sample $(X_1,\ldots,X_n)$ and the associated sample of ascending order statistics $(X_{1:n} \leq \cdots \leq X_{n:n})$, the classical Hill estimator of a positive EVI is the average of the k log-excesses $V_{ik} := \ln X_{n-i+1:n} - \ln X_{n-k:n}$, $1 \leq i \leq k < n$. The aforementioned class, which generalizes the Hill estimator, comes from the Lehmer mean of order p of k positive numbers and is defined as $L_p(k) := 1/p\left(\sum_{i=1}^{k} V_{ik}^{p} / \sum_{i=1}^{k} V_{ik}^{p-1}\right)$. The asymptotic behaviour of the L_p−EVI-estimators has revealed very nice results in the sense of minimization of the mean square error, at optimal levels. However, for finite samples the estimates show the usual trade-off between bias and variance, depending on k. Besides k there is also the need of the choice of p. Bootstrap methodology has revealed to be particularly promising in the estimation of parameters of extremes events. A bootstrap algorithm for an adaptive estimation of the tuning parameters in $L_p(k)$ is given, allowing a reliable EVI-estimation.

C0352: **Estimation of the linear fractional stable motion: Numerical techniques and results**
Presenter: **Dmitry Otryakhin**, Aarhus University, Denmark
Co-authors: Mark Podolskij, Stepan Mazur

Linear fractional stable motion is a non-Gaussian analogue of fractional Brownian motion which is used in some areas of physics as well as in network traffic modeling. Several parameter estimation techniques for this type of motions have been studied, while some of the estimators were developed with the corresponding limit theory. The main numerical features of the latter ones are shown. On top of that, the ideas behind the architecture of R package rlfsm, developed for numerical studies of lfsm, are discussed with possible extensions to applications in Levy-driven processes.

CG005 **Room Orion Hall** ADVANCES IN COMPUTATIONAL ECONOMETRICS **Chair: Massimiliano Caporin**

C0198: **GMM for regression models with exogenous regressors and non-spherical disturbances**
Presenter: **Taku Yamamoto**, Hitotsubashi University, Japan
Co-authors: Hiroaki Chigira

GMM (generalized method of moments) is applied to regression models with the endogeniety probrem where regressors are correlated with disturbances. When there is no endogeniety, GMM is usually not recommended in econometrics textbooks, since GMM reduces to OLS (ordinary least squares) when the original regressors themselves are used as its instrumental variables. The purpose is to present the effectiveness of GMM for linear regression models of exogenous regressors with non-spherical disturbances, that is, the disturbances are heteroscadastic and/or serially correlated. It appears to be an overlooked feature of GMM. We in particular demonstrate that GMM with the suitable extra instrumental variables in addition to the original regressors can improve efficiency of the estimator when disturbances are non-spherical. For the model with heteroscadastic disturbances, we propose the extra instrumental variables which are modifications of those proposed for PGLS (partial generalized least squares) by Amemiya. For the model with autocorrelated disturbances, we propose lagged regressors as the extra instrumental variables. The analytical results for some illustrative models and the suitably designed Monte Carlo experiments exhibit that GMM with these extra instrumental variables gives more efficient estimates than OLS.

C0342: **Discretization of the tail density function and adjusted evaluation measures**
Presenter: **George-Jason Siouris**, University of the Aegean, Greece
Co-authors: Alexandros Karagrigoriou, Ilia Vonta, Despoina Skilogianni

After extensive investigation on the statistical properties of financial returns, three properties have shown to be present in most, if not all, financial returns. Their existence has been the source of most problems associated with the estimation of the underlying risk of assets. These are often called the three stylized facts of financial returns and are volatility clusters, fat tails and nonlinear dependence. In order to forecast the asset volatility, a number of different models have been developed over the years. Each of them offers an answer on a specific aspect of the problem at hand. Many of these models incorporate skewed, fat-tailed distributions. The disadvantage of this approach, is that even with the simple and well-known Student distribution closed-form expected shortfall expressions are not available. This is also the case for many asymmetric heavy-tailed distributions. We propose a discretization of the tail density function which is a logical approach for resolving the above issues, since the nature of returns is discrete, as the market always operates on a specific accuracy. As a result, with the use of adjusted evaluation measures we provide improved expected percentage shortfall estimations. Illustrative examples verify the advantages of the proposed methodology.

C0358: **Tests of cointegration rank with strong persistence and heavy-tailed errors**
Presenter: **Niklas Ahlgren**, Hanken School of Economics, Finland
Co-authors: Paul Catani

Financial time series have several distinguishing features which are of concern in tests of cointegration. An example is testing the approximate non-arbitrage relation between the credit default swap (CDS) price and bond spread. Strong persistence and very high persistence in volatility are stylised features of cointegrated systems of CDS prices and bond spreads. It is shown that tests of cointegration rank in the heteroskedastic vector autoregressive model have low power under such conditions. Obtaining high power requires more than 1000 observations. Hill estimates of the tail index indicate that the distribution of the errors has heavy tails with finite variance but infinite fourth moment. Asymptotic and bootstrap tests of cointegration rank are unreliable if the errors are heavy-tailed with infinite fourth moment. Monte Carlo simulations indicate that the wild bootstrap (WB) test may be justified with heavy-tailed errors which do not have finite fourth moment. The tests are applied to daily observations from 2010 to 2016 on the CDS price and bond spread of US and European investment-grade firms. The WB test accepts cointegration for most firms in the full sample period. The evidence for cointegration is weak in sub-sample periods.

C0384: **On a novel spherical Monte Carlo method via group representation**
Presenter: **Huei-Wen Teng**, National Chiao Tung University, Taiwan
Co-authors: Ming-Hsuan Kang, Runze Li

Accurate and efficient calculation of d-dimensional integrals for large d is of crucial importance in various scientific disciplines. Via spherical transformation, standard spherical Monte Carlo estimators consist of independent radii and a set of unit vectors uniformly distributed on a unit sphere. A random orthogonal group is used to rotate a set of unit vectors simultaneously, and can be generated by applying the Gram-Schmit procedure to a $d \times d$ matrix with i.i.d. standard normal random variables as entries. The generation of a random orthogonal group is however computationally demanding. To overcome this problem, a novel spherical Monte Carlo approach is proposed via group representation: By constructing a subgroup of the orthogonal groups, the spherical integral is calculated using the group orbit of a random unit vector. In this case, the generation of a random unit vector only needs d i.i.d. standard normal random variable. The proposed method outperforms existing methods in terms of computation efficiency in high-dimensional cases. Theoretical properties of the proposed subset are provided. Extensive numerical experiments with applications in finance confirm our claims.

CG013 **Room Vega Hall** ADVANCES IN ROBUST STATISTICS **Chair: Matias Salibian-Barrera**

C0183: **Testing marginal homogeneity for paired ordinal data: A robust likelihood approach**
Presenter: **Tsung-Shan Tsou**, Institute of Statistics, National Central University, Taiwan, Taiwan

Pairing lessens heterogeneity between subjects but introduces correlation that is nuisance to our interest. This makes modeling the underlying probability structure more complex and inference more challenging. The problem is more severe especially for discrete data. Instead of confronting the nuisance parameter, we use the simpler model for the parallel design and develop from it a robust score statistic for testing the marginal homogeneity of two distributions of paired ordinal data. The effect of the overlooked correlation is recuperated externally from data. We provide simulations and real data analyses to demonstrate the effectiveness of the robust test.

C0204: **Informative transformation of responses that can be positive or negative**
Presenter: **Anthony Atkinson**, London School of Economics, United Kingdom
Co-authors: Marco Riani

The parametric family of power transformations to approximate normality analysed by Box and Cox can be applied only to positive data. This transformation has been generalized to allow for the inclusion of zero and negative response values, which arise for example in data on GNP growth and company profits and in the differences in measurements before and after treatment. The aim is to describe the use of constructed variables to provide an approximate score statistic for the transformation which avoids the numerical optimization required for estimation of the transformation parameter using maximum likelihood. The resulting statistic is based on aggregate properties of the data. Robust analysis of the data with the forward search provides a series of subsets of the data of increasing size, ordered by closeness to the fitted model for each subset size. The "fan plot" of the statistics for these subsets against subset size clearly indicates the effect of individual observations, especially outliers, on the estimated transformation parameter. The score test is extended to determine whether positive or negative observations require different transformations, leading to an informative extended fan plot. The methods will be illustrated with several examples, one from Darwin on cross- and self-fertilized plants. There will be some discussion of the distributions of the test statistics.

C0340: **Robust parsimonious approach for model based clustering**
Presenter: **Agustin Mayo-Iscar**, Universidad de Valladolid, Spain
Co-authors: Luis Angel Garcia-Escudero, Marco Riani, Andrea Cerioli

Trimmed k-means were introduced 20 years ago for robustifying the well-known k-means. The base of the success of this simple procedure was the joint application of impartial trimming and constraints. Later, TCLUST approaches extended this procedure by reducing the strength of the constraints, implicit in the application of that procedure, in order to fit better the existing patterns in data sets from mixtures of normal multivariate distributions. Impartial trimming application allows us to avoid the undesired effect that deviations from the assumed model produce in maximum likelihood based estimators. Constraints are useful for getting a well posed estimation problem and for reducing the prevalence of spurious local maximizers. Trimming and constraints based procedures were also designed for robustifying the estimation of clusters around linear subspaces and the estimation of the mixture of factor analyzers model. TCLUST methodologies are available in the TCLUST package in CRAN and in the FSDA library in MATLAB. Now we are developing robust estimators for a parsimonious collection of models for being incorporated in these packages. As usual, a BIC application allows to select the model. It will allow the users to get adaptive robust estimations for their data sets.

C0387: **Fast computation of Tukey trimmed regions and median in dimension** $p > 2$
Presenter: **Pavlo Mozharovskyi**, CREST-Ensai, France
Co-authors: Xiaohui Liu, Karl Mosler

Given data in $\mathbb{R}^p$, a Tukey κ-trimmed region is the set of all points that have at least Tukey depth κ w.r.t. the data. As they are visual, affine equivariant and robust, Tukey regions are useful tools in nonparametric multivariate analysis. While these regions are easily defined and interpreted, their practical use in applications has been impeded so far by the lack of efficient computational procedures in dimension $p > 2$. We construct two novel algorithms to compute a Tukey κ-trimmed region, a naïve one and a more sophisticated one that is much faster than known algorithms. Further, a strict bound on the number of facets of a Tukey region is derived. In a large simulation study the novel fast algorithm is compared with the naïve one, which is slower and by construction exact, yielding in every case the same correct results. Finally, the approach is extended to an algorithm that calculates the innermost Tukey region and its barycenter, the Tukey median.

Wednesday 29.08.2018 09:00 - 10:30 Parallel Session E – COMPSTAT2018

CI006 Room Cuza Hall NEW STATISTICAL DEVELOPMENTS AND INSIGHTS VIA THE STEIN METHOD Chair: Christophe Ley

C0221: **A Stein variational framework for deep probabilistic modeling**

Presenter: **Qiang Liu**, UT Austin, United States

Modern AI and machine learning techniques increasingly depend on highly complex, hierarchical (deep) probabilistic models to reason with complex relations, and make decisions under uncertain environment. This, however, casts a significant demand on developing efficient computational methods for highly complex probabilistic models in which exact calculation is prohibitive. We discuss a new framework for approximate learning and inference that combines ideas from Stein's method, an advantaged theoretical technique developed by mathematical statistician Charles Stein, with practical machine learning and statistical computation techniques such as variational inference, Monte Carlo, optimal transport and reproducing kernel Hilbert space (RKHS). Our framework provides a new foundation for probabilistic learning and reasoning and allows us to develop a host of new algorithms for a variety of challenging learning and AI tasks, that are significantly different from, and have critical advantages over, traditional methods. Examples of applications include computationally tractable goodness-of-fit tests for evaluating highly complex models, scalable Bayesian computation, deep generative models, and sample efficient policy gradient for deep reinforcement learning.

C0213: **Differential Stein operators for multivariate distributions and applications**

Presenter: **Yvik Swan**, Universite de Liege, Belgium

Co-authors: Gesine Reinert, Guillaume Mijoule

After introducing the concept of "directional Stein operator", we discuss several types of gradient and divergence based differential Stein operators for multivariate random vectors with absolutely continuous densities p on $\mathbb{R}^d$, $d \geq 1$. In particular, we provide minimal conditions on p to guarantee the existence of a score function and a Stein kernel; this leads to probabilistic integration by parts formulas which generalize Stein's famous Gaussian covariance lemma. We illustrate the operators and identities on the family of elliptical distributions (particularly the Gaussian, multivariate Student, and power exponential distributions), hereby providing new tractable operators which moreover bear nice interpretations. We introduce a new family of kernelized Stein discrepancies. Several applications are outlined: aside from the habitual Stein-type measures of discrepancies, we also discuss problems of goodness-of-fit testing.

C0205: **A probabilistic measure of the impact of the prior in Bayesian statistics**

Presenter: **Christophe Ley**, Ghent University, Belgium

Co-authors: Gesine Reinert, Yvik Swan, Fatemeh Ghaderinezhad

A key question in Bayesian analysis is the effect of the prior on the posterior, and how this effect could be assessed. As more and more data are collected, will the posterior distributions derived with different priors be very similar? This question has a long history. It is well-known that, asymptotically, the effect of the prior wanes as the sample size tends to infinity. We are interested, at fixed sample size, in explicit bounds on some measure of the distributional distance between posteriors based on a given prior and the no-prior data-only based posterior, allowing us to detect the effect of the prior at fixed sample size. To reach this goal, we have developed new aspects of the celebrated Stein Method for asymptotic approximations. Besides theoretical results we will also show numerical computations that illustrate how to measure a prior's impact.

CO107 Room Clio Hall ARS-IASC SESSION: DEPENDENCE VIA COVARIANCES AND SPATIAL STRUCTURES Chair: Min-ge Xie

C0193: **Optimal Bayesian minimax rates for unconstrained large covariance matrices**

Presenter: **Jaeyong Lee**, Seoul National University, Korea, South

Co-authors: Kyoungjae Lee

The optimal Bayesian minimax rate for the unconstrained large covariance matrix of multivariate normal sample with mean zero is obtained when both the sample size, n, and the dimension, p, of the covariance matrix tend to infinity. Traditionally the posterior convergence rate is used to compare the frequentist asymptotic performance of priors, but defining the optimality with it is elusive. We propose a new decision theoretic framework for prior selection and define Bayesian minimax rate. Under the proposed framework, we obtain the optimal Bayesian minimax rate for the spectral norm for all rates of p. We also considered Frobenius norm, Bregman divergence and squared log-determinant loss and obtain the optimal Bayesian minimax rate under certain rate conditions on p. A simulation study is conducted to support the theoretical results.

C0200: **A variational approach to a nonparametric density estimation using dependent species sampling models**

Presenter: **Seongil Jo**, Chonbuk National University, Korea, South

The dependent species sampling model is a Bayesian nonparametric model for dependent data which estimates the probability density functions indexed by time or space. The dependent species sampling model is often applied a large data set, but the posterior sampling algorithm with Markov chain Monte Carlo can require lengthy computation time. We propose a variational method for the posterior approximation. We check the accuracy and speed of the variational method by simulation studies and illustrate the proposed method with some real data sets.

C0291: **A semiparametric mixture method for local false discovery rate estimation**

Presenter: **Woncheol Jang**, Seoul National University, Korea, South

A two-component semiparametric mixture model is proposed to estimate local false discovery rates in multiple testing problems. The two pillars of the proposed approach are Efron's empirical null principle and log-concave density estimation for the alternative distribution. Our method outperforms other existing methods, in particular when the proportion of null is not that high. It is robust against the misspecification of alternative distribution. A unique feature of our method is that it can be extended to compute the local false discovery rates by combining multiple lists of p-values. We demonstrate the strengths of the proposed method by simulation and several case studies.

C0294: **High-dimensional Markowitz's portfolio optimization problem: An empirical comparison of covariance matrix estimators**

Presenter: **Johan Lim**, Seoul National University, Korea, South

Co-authors: Young-Geun Choi, Sujung Choi

The performance of recently developed regularized covariance matrix estimators for Markowitz's portfolio optimization, and of the minimum variance portfolio (MVP) problem in particular, is compared. We focus on seven estimators that are applied to the MVP problem in the literature; three regularize the eigenvalues of the sample covariance matrix, and the other four assume the sparsity of the true covariance matrix or its inverse. Comparisons are made with two sets of long-term S&P 500 stock return data that represent two extreme scenarios of active and passive management. The results show that the MVPs with sparse covariance estimators have high Sharpe ratios but that the naive diversification (also known as the uniform -on market share- portfolio) still performs well in terms of wealth growth.

CO032 Room Orion Hall COMPUTATIONAL STATISTICS FOR APPLICATIONS Chair: Marta Disegna

C0260: **A new method for analyzing ethnic mixing: Studies from Southern California**

Presenter: **Madalina Olteanu**, Pantheon-Sorbonne University, France

Co-authors: William Clark, Julien Randon-Furling

An ongoing question in studies of residential segregation is how best to capture the complexity evident in multi-ethnic cities and in cities with growing immigrant populations. One of the difficult issues is to capture local complexity and to visualize how that complexity changes over space. Using a new mathematical framework based on trajectories of aggregated spatial units, one gets a flexible method for capturing segregation as a multiscalar phenomenon. Thus, the key to the analysis is studying how far, in terms of distribution distances for example, any neighborhood is from the city wide measure of ethnicity. We use these methods to investigate social mixing in the Southern California metropolitan area. We find that these methods provide excellent measures of the patterns of mixing across urban space and that the trajectories reveal the spatial speed at which the process of convergence takes place. From the studies of Southern California and Los Angeles, we show how relative isolation generates "hot spots" of slow convergence to region wide averages. The advance in measuring segregation with the trajectory convergence analysis is that we have both numerical measures of the level of segregation and a visual picture of the outcomes of social distance.

C0405: **A sparse tensor subspace method for identifying biological modulators based on multilayer gene network analysis**
Presenter: **Heewon Park**, Yamaguchi University, Japan
Co-authors: Rui Yamaguchi, Seiya Imoto, Satoru Miyano

Identifying crucial biological modulators has drawn a large amount of attention in precision medicine to understand molecular-cellular characteristic of disease. To identify biological modulators (e.g., candidate anti-cancer drug), we consider tensor similarity test for significant modulator-specific characteristic of disease. We consider drug sensitive and resistant gene network tensors, and then measure distance between the two tensors on tensor subspace. Although, the high dimensional genomic data include noisy features inevitably and the noise disturbs the process of not only similarity test but also subspace identification, relative little attention was paid to incorporating sparsity into tensor subspace method. In order to efficiently construct drug sensitivity-specific tensor subspace, we propose a novel sparse common component analysis based on L1-type regularization. By incorporating sparsity into subspace identification, our method constructs tensor subspace based only on the crucial common edges of multilayer gene networks without disturbance of noise. Thus, we can effectively extract characteristic of gene regulatory system in drug sensitive and resistant cell lines. We then propose a statistical test based on the similarity measure of tensors on the constructed tensor subspace to identify biological modules. The proposed method is applied to identify candidate anti-cancer drugs based on drug sensitive and resistance gene network tensors.

C0212: **Compositional analysis in tourism**
Presenter: **Berta Ferrer-Rosell**, Universitat de Lleida, Spain
Co-authors: Germa Coenders

Compositional Data analysis (CoDa) is the standard statistical method when data contain information about the relative importance of parts of a whole, typically but not necessarily with a fixed sum. Many research questions in the field of tourism have to do either with distribution of a whole (e.g., share or allocation), or with relative importance (e.g., dominance, concentration, profile, etc.). Some examples of such research questions might be: What are the determinants of market share of a given destination or tourism product? Or, which origins and destinations concentrate the most tourist flows, per season or tourist segment? The main purpose is to present the manner in which CoDa solves statistical problems that arise (e.g. spurious correlations) when treating compositional data with classical statistical methods, as well as, to show how to apply the most common exploratory tools to analyse compositions (data transformations, distances, CoDa biplot and cluster) by means of real examples.

C0274: **Fuzzy clustering with spatial-time information**
Presenter: **Marta Disegna**, Faculty of Management, Bournemouth University, United Kingdom
Co-authors: Pierpaolo Durso, Riccardo Massari

Clustering geographical units based on a set of quantitative features observed at several time occasions requires dealing with the complexity of both space and time information. In particular, one should consider (1) the spatial nature of the units to be clustered, (2) the characteristics of the space of multivariate time trajectories, and (3) the uncertainty related to the assignment of a geographical unit to a given cluster on the basis of the above complex features. Existing spatial-time clustering models can be distinguished into non-spatial time series clustering based on a spatial dissimilarity measure; spatially constrained time series clustering; density-based clustering; model-based clustering. The aim is to discuss a novel spatially constrained multivariate time series clustering for units characterised by different levels of spatial contiguity. In particular, the Fuzzy Partitioning Around Medoids algorithm with Dynamic Time Warping distance and spatial penalization terms is applied to classify multivariate time series. This clustering algorithm has been theoretically presented and discussed through different simulation examples, highlighting its main advantages. A real case study has been presented to illustrate the usefulness and effectiveness of the suggested clustering method for tourism spatial-time series, especially in the identification of the spatial tourism spill-over effect.

CO072 Room Vega Hall DATA SCIENCE AND CLASSIFICATION — Chair: Patrick Groenen

C0316: **Biplot perspectives on student performance at the Copperbelt University in Zambia**
Presenter: **Niel Le Roux**, Stellenbosch University, South Africa
Co-authors: Mwanabute Ngoy

Education in general, and tertiary education in particular are the engines for sustained development of a nation. This case study considers the vital role of the Copperbelt University (CBU) in Zambia in delivering the necessary knowledge and skills requirements for the development of Zambia and the neighbouring southern African region. Of importance is thus to investigate relationships between school and university results at the CBU. Biplot representations of student performance at the CBU for the period 2000 to 2013 reveal changes in cut-off values for university entrance resulting in changes in the performance of the student body; students were achieving higher scores at school level which could not translate necessarily into higher academic achievement at university; certain school subjects are better indicators of university performance; policies of making school results available as grades rather than actual percentages can have a marked influence on expected university achievements. Furthermore, most school variables had limited discrimination power to differentiate between successful and unsuccessful students. In particular, it is shown how categorical principal component analysis and categorical canonical variate analysis biplots differentiate between different groups of university performers.

C0322: **A naive visualisation of data science**
Presenter: **Sugnet Lubbe**, Stellenbosch University, South Africa

A graphical representation is constructed in order to visualise the different aspects and concepts related to Data Science. The popular search engine Google uses Data Science to customise our search according to our historical search data. DuckDuckGo, an internet search engine that emphasizes protecting searchers' privacy and avoiding the filter bubble of personalized search results was utilised to obtain a hopefully unbiased result of the question "What is Data Science?" A total of 47 sites' descriptions were extracted and the key words tabulated. The use of Data Science in constructing a visualisation of "What is Data Science" will be illustrated utilising different forms of multidimensional scaling.

C0381: **Using clustered heat maps to improve the selection of a clustering algorithm and its parametrization**
Presenter: **Leonardo Feltrin**, Laurentian University, Canada
Co-authors: Martina Bertelli

Cluster analysis is a discipline that aims at finding groups of similar entities in a data set. One of the outstanding problems of cluster analysis is the selection of an appropriate granularity and clustering algorithm. Many solutions attempt to evaluate the clustering quality to select an adequate cluster number (K) and an optimal classifier. Proposed strategies attempt to locate automatically a set of threshold values to optimize the position of the decision boundaries, causing often unwanted information loss. Interpretive solutions based on external validation measures of cluster quality can be misleading since they provide quality indices that depend upon the validation strategy and are difficult to interpret, limiting the capacity of evaluating the partitioning process. More exhaustive information is needed to permit embedding of domain knowledge to improve the classification outcome. A dynamic, computational workflow was designed to obtain n-dimensional visualizations of the clustering process (based on Clustered Heat Maps with custom annotations). Synthetic data experiments show that this workflow facilitates the selection of an appropriate granularity level, making more explicit the results of multiple clustering algorithms and relative parametrizations. This approach exposes some of the weaknesses of external cluster validation methods.

C0311: **Computing and validating neural reliability from EEG recordings**
Presenter: **Pieter Schoonees**, Erasmus University Rotterdam, Netherlands
Co-authors: Niel Le Roux

Evidence has emerged from the neuroscience literature that the level of intersubject synchronization between the neural responses of different subjects to, for example, movies is related to population-level measures of movie success, such as box office performance. Measures of such intersubject similarity are also known as neural reliability measures. The assumption is that the more engaging a naturalistic stimuli such as a movie is, the more similar the responses are even when comparing across subjects. Several studies have shown this empirically, using a variety of methods including correlation-based distance measures and component analysis techniques similar to canonical correlation analysis. We discuss these approaches and how to validate them using simulated data.

CG089 **Room Mezzanine Lounge** **HIGH-DIMENSIONAL AND NONPARAMETRIC STATISTICS** **Chair: Johannes Lederer**

C0339: **A sparse variable selection approach in multiscale local polynomial density estimation**
Presenter: **Maarten Jansen**, ULB Brussels, Belgium

The multiscale local polynomial transform (MLPT, implemented in the Matlab Wavelet Toolbox) is a slightly overcomplete data representation that combines the sparsity of a wavelet decomposition and the smoothness properties of a local polynomial smoothing procedure on nonequispaced data points. The MLPT adopts the bandwidths as user controlled nondyadic resolution levels. Careful application of the MLPT enables us to perform density estimation without preprocessing and corresponding possible loss of information. The densities under consideration may have multiple singularities at unknown locations. The presence of singularities, as well as the intermittent nature of the density estimation problem itself, with intervals of low and high intensities, are natural arguments for a multiscale approach. Moreover, taking the sample values as data points, leads immediately to a nonequispaced problem. The MLPT basis functions can be used in the design matrix of a sparse high-dimensional regression model, with asymptotically exponential responses, where the responses are given by the spacings, i.e., the differences between successive values in the ordered sample. We explain how the Karush-Kuhn-Tucker conditions for the L1-regularised exponential regression model can be approximately solved by soft-thresholding the MLPT applied to the inverse spacings. Optimal thresholds can be chosen by (estimated) minimisation of the Kullback-Leibler distance.

C0389: **Simultaneous test for mean vectors and covariance matrices among k populations for high-dimensional data**
Presenter: **Takahiro Nishiyama**, Senshu University, Japan
Co-authors: Hayate Ogawa, Masashi Hyodo

A simultaneous test for mean vectors and covariance matrices among k populations in non-normal high-dimensional data is proposed. Since the classical hypothesis testing methods based on the likelihood ratio degenerate when the dimensionality exceeds the sample size, we propose a new L^2-norm-based test. To construct a test procedure, we propose a test statistic based on both an unbiased estimator of differences of mean vectors and covariance matrices. Also, we derive an asymptotic null distribution of this test statistic. Finally, we study the finite sample and dimension performance of this test via Monte Carlo simulations. We demonstrate the relevance and benefits of the proposed approach for some alternative mean and covariance structures.

C0402: **A fast algorithm for univariate log-concave density estimation**
Presenter: **Yong Wang**, University of Auckland, New Zealand

A new fast algorithm is proposed and studied for computing the nonparametric maximum likelihood estimate of a univariate log-concave density. In each iteration, the newly extended algorithm includes, if necessary, new knots in aid of a gradient function, renews the changes of slope at all knots via a quadratically convergent method and removes the knots at which the changes of slope become zero. Theoretically, the characterisation of the nonparametric maximum likelihood estimate is studied and the algorithm is guaranteed to converge to the unique maximum likelihood estimate. Numerical studies show that it outperforms other algorithms that are available in the literature. Applications to some real-world financial data are also given.

C0435: **Flexible shrinkage of large-dimensional covariance matrices**
Presenter: **Nicolas Tavernier**, KU Leuven, Belgium
Co-authors: Geert Dhaene

An optimal rule is derived for shrinking large-dimensional sample covariance matrices under Frobenius loss. The rule generalizes the optimal linear shrinkage rule to broader parametric families of rules. The families include, for instance, polynomial and spline rules. The oracle version of the optimal rule is very simple and attains the lower bound on the Frobenius loss in finite samples. A feasible version is proposed and approximates the lower bound under large-dimensional asymptotics where $p/n \to c > 0$. In a variety of settings, nonlinear shrinkage is found to substantially improve upon linear shrinkage and to perform on par with the current state-of-the-art, but highly complex, nonlinear shrinkage estimator.

CP001 **Room Cocktail Hall** **POSTER SESSION I** **Chair: Panagiotis Paoullis**

C0187: **Lifestyle and dietary habits of community-dwelling elderly: A comparative study between a town and a city**
Presenter: **Chisako Yamamoto**, Shonan University of Medical Sciences, Japan

The aim is to clarify differences of lifestyle and dietary habits and gender differences between a town (Town) and a city (City) elderly. Data were collected in 2004 in both Town and City. Analysis subjects were 1,538 and 13,182 men and women in Town and City, respectively. According to intellectual activity scores, they were classified into people with dementia (PWD), people with probable dementia (PPD) and cognitively intact people (CIP). The chi-square, Kruskal-Wallis, Mann-Whitney U and Bonferroni's multiple comparison tests were performed. Lifestyle items were smoking, alcohol drinking, exercise, sleep duration per day, going out, daytime lying duration in bed and breakfast eating and dietary items meat/poultry, soy products, eggs, oily fish, dairy products, fruits, vegetables and others. The tests revealed statistically significant differences between City CIP and PPD/PWD women in most dietary items and between City CIP and PPD men and between Town CIP and PPD women in some items, while no differences in dietary items were observed in Town men. In lifestyle items, except smoking, significant differences were shown in both Town and City. The conclusion is that the health behavior of women, especially in dietary habits, was better than that of men, while that of City men was better than that Town men.

C0249: **Nonlinear and kernel regression methods for interval-valued data**
Presenter: **Kee-Hoon Kang**, Hankuk University of Foreign Studies, Korea, South
Co-authors: Jeong-Taek Jang

Symbolic data are difficult to represent by single value because each observation object has internal structure and variation. Interval data, which is one of these symbolic data, is given as an interval in which all observation objects are not single values. We introduce a regression model using kernel functions and a method of fitting nonlinear regression models to the interval data. We also propose to apply the local linear regression model using the kernel function to the interval data analysis. Various simulations are carried out according to the distribution of the center point and the range by using each method. When various conditions are considered, it is confirmed that the performance of the model based on the proposed local linear regression is quite good. Also, it can be seen that the proposed method shows better performance in the situation where the nonlinear regression function is hard to be estimated well in the real data analysis.

C0335: **Determining the number of components of a PLS regression on incomplete data**

Presenter: **Titin Agustin Nengsih**, University of Strasbourg, France

Co-authors: Frederic Bertrand, Myriam Maumy-Bertrand, Nicolas Meyer

Missing data is known to be a concern for the applied researcher. Several methods have been developed for handling incomplete data. Imputation is the process of substituting missing data before estimating the relevant model parameters. PLS regression is a multivariate model estimated either by the SIMPLS or NIPALS algorithm. The goal is to analyze the impact of the missing data proportion on the estimation of the number of components of a PLS regression by simulations. We compare the criteria for selection of the number of components of a PLS regression on incomplete data and PLS regression on imputed data set which used three methods of imputation: multiple imputations by chained equations (MICE), k-nearest neighbors imputation (KNNimpute) and a singular value decomposition imputation (SVDimpute). The compared criteria are Q2-LOO, Q2-10 fold, AIC, AIC-DoF, BIC and BIC-DoF on different proportions of missing data (from 1 to 50%) and under a MCAR assumption and a MAR assumption. The results show that MICE had the closest to the correct number of components at each frequency of missingness although it needs a long time for the execution. Furthermore, NIPALS-PLSR ranked second, followed by KNNimpute and SVDimpute. Whatever the criterion, except Q2-LOO, the number of components in a PLS regression is far from the true one and tolerance to incomplete data sets depends on the sample size, the proportion of missing data and the chosen component selection method.

C0366: **Explanation of the importance of medication adherence in antiretroviral therapy by using random numbers**

Presenter: **Shinobu Tatsunami**, St. Marianna University School of Medicine, Japan

Co-authors: Takahiko Ueno

Even after remarkable developments in an antiretroviral therapy, almost perfect adherence to medication intake is needed in patients. Incomplete adherence may result in the failure of a therapy. However, the frequency of failure is extremely low under good control by medical experts. Therefore, there is little statistical data that can make patients realize the potential risk of incomplete adherence. In this context, a simple dynamic model that expresses the failure of therapy by using random numbers has been developed. This model contains three main variables of viral concentration $v(t)$, strength of the immune activity $L(v)$, and the concentration of a drug $c(t)$. The most fundamental assumption is that the probability of the appearance of a drug-resistant virus depends on the time-derivative of viral concentration, dv(t)/dt. An additional assumption is that the viral concentration v(t) does not reach zero but maintains a small value such as the attainable limit. Under this assumption, the dynamic equation could describe the divergence of viral concentration with a low frequency. The frequency of viral divergence depended on the various patterns of incompleteness in medication adherence. However, it showed non-sensitive dependence on the attainable limit of viral concentration.

C0371: **A note on the analysis of early stage breeding experiments**

Presenter: **Stanislaw Mejza**, Poznan University of Life Sciences, Poland

Co-authors: Iwona Mejza

In plant breeding trials, during the early stages of the improvement process, it is not possible to use an experimental design that satisfies the requirement of replicating all the treatments, because of the large number of genotypes involved, the small amount of seed and the low availability of resources. Hence, unreplicated designs are used for early generation testing when hundreds or even thousands of new genotypes are to be evaluated in the same trial using a limited amount of seed that is enough for one replication only. To control the real or potential heterogeneity of experimental units, control (check) plots are arranged in the trial. There are many methods of using the information resulting from check plots. The main tool for exploring this information will be based on a response surface methodology. To begin with we attempt to identify the response surface characterizing the experimental environments. The obtained response surface is then used to adjust the observations for genotypes. Finally, the adjusted data are used for inferences concerning the next stages of the breeding program. The theoretical considerations are illustrated with an example involving spring barley.

C0375: **Development and application of data providing and visualization system for EMS**

Presenter: **Kazuki Konda**, Graduate school of Tokai University, Japan

Co-authors: Yoshiro Yamamoto, Hideaki Takenaka, Hideki Kimura, Kota Fukuda, Takashi Nakajima

Energy management systems (EMSs) that use meteorological data are considered. Many people avoid using meteorological data because the format is not standardized, and so we investigated a way to simplify the data for the user. In particular, we simplified the way in which the data is provided to a system that estimates solar radiation. This system uses high-definition data obtained from the Himawari 8 geostationary meteorological satellite. We built data provision and visualization system for EMS on Microsoft cloud Azure and On-premise Linux server. We explain the advantages and inconveniences of using the cloud environment. We also report on the development of a visualization system to support solar car race (WSC2017) as an application.

C0423: **Novel standardization methods for preprocessing multivariate data used in predictive modeling**

Presenter: **Emily Grisanti**, TU Freiberg, Germany

Co-authors: Matthias Otto

An essential part of multivariate analysis is the preprocessing of the data set, since most of the predictive models assume data to be standardized. This is why the Standard Normal Variate (SNV) transformation, which subtracts the mean value and divides it by the standard deviation sample by sample, has become very popular. For highly correlated data which are often found in the context of analytical measurement, e.g. vibrational spectroscopy, performing the SNV over the full spectral range is in some cases not sufficient for removing unwanted effects, e.g. influence of stray light. Three different standardization methods are presented, that apply SNV to defined sequential windows rather than to the full spectrum: Dynamic Localized SNV (DLSNV), Peak SNV (PSNV) and Partial Peak SNV (PPSNV). DLSNV is an enhancement of the Localized SNV (LSNV), which allows a dynamic starting point of the localized windows on which the SNV is executed individually. Peak and Partial Peak SNV are based on the selection of ranges from the spectra that have a high correlation to the target value and perform SNV on these essential windows. The prediction errors of two regression models in chemical analytics are shown to be reduced by up to 16% and 29%, respectively, compared to LSNV.

C0425: **A note on the computational complexity of the sample variance over an interval-valued dataset**

Presenter: **Ondrej Sokol**, University of Economics, Prague, Czech Republic

Co-authors: Miroslav Rada

Assume an interval-valued dataset. The exact values of data are not known; however, the intervals which these values belong to are observable. This is a common situation when we work with censored, categorized or rounded data. A similar problem also occurs when we work with measurement error or predicted values. In these situations, the computation of the identification region of even basic statistics can be computationally expensive. We focus on the problem of computing the maximal sample variance over interval data, which can be expressed maximization of a convex function over a convex set. The problem is known to be NP-hard. However, various polynomial-time algorithms were constructed for special cases. One of the most interesting is an algorithm usable in most of common cases. It works in polynomial time in k, where k is the maximal number of *narrowed* intervals intersecting at one point; narrowed means that the intervals are shrinked proportionally to the size n of the dataset, namely to $1/n$ of their original width. Considering randomly generated datasets, our experiments allowed for conjecturing that k is at most of logarithmic size for a reasonable choice of the data-generating process. This implies that the algorithm works in polynomial-time in average. We discuss the computational complexity from the view of randomly generated data and its relation to other problems in statistics such as so called *birthday problem*.

C0428: **A survey of recent statistical methods for EEG-FMRI**

Presenter: **Mohammad Fayaz**, Shahid Beheshti University of Medical Sciences, Iran

Statistical analysis of EEG-FMRI in neuroscience have major challenges both from theoretical and practical aspects such as complex relationships, sparsity and smoothing. In this manner, many Functional Data Analysis methods are developed which considered function or curve as building blocks in models. But Functional data analysis have different frameworks with different basis functions. The aim is to review recent developments in statistical methods specially functional data analysis methods for EEG-FMRI.

C0440: **An extreme value analysis of top performing UK winter wheat producers**

Presenter: **Emily Mitchell**, The University of Nottingham, United Kingdom

Co-authors: Gilles Stupfler, Andrew Wood, Neil Crout, Paul Wilson

Using the responses to a UK-based survey, we present the first application of extreme value theory in an agricultural setting to complement the previous studies conducted from a classical central perspective in this field. The Farm Business Survey collects a substantial amount of information annually from farms across England and Wales with the purpose of providing farmers with an overview of farming performances. Winter wheat is the most popular crop grown in the UK due to its optimal growing conditions; therefore, we focus on winter wheat production from 2006 to 2015 and extract a subset of variables from this data set, among which the obtained yield and net margin, and apply a number of established extreme value analysis methods. In particular, we use a mix of Peaks-Over-Threshold and semi-parametric approaches to fit distributions to the tail before ultimately producing extreme quantile estimates. We conclude by stratifying about pesticide usage for top UK winter wheat producers to address the inconsistencies of management practices; moreover, we compare quantile estimates between stratum and discuss the implications of our results.

C0233: **Bayesian modeling of individual growth variability using back-calculation: Application to pink cusk-eel**

Presenter: **Freddy Omar Lopez Quintero**, Telefonica-UTFSM-UV-PUCV, Chile

The von Bertalanffy growth function with random effects has been widely used to estimate growth parameters incorporating individual variability of length at age. Inferred trajectories of individual growth can be assessed from growth marks in hard body parts such as otoliths using either mark-recapture or back-calculation of length-at-age. We combine recent studies in non-Gaussian distributions and a Bayesian approach to model growth variability using back-calculated data in harvested fish populations. We presumed that errors in the VBGF can be assumed as a Student-t distribution, given the abundance of individuals with extreme length values. The proposed method was applied and compared to the standard methods using back-calculated length-at-age data for pink cusk-eel (*Genypterus blacodes*) off Chile. Considering several information criteria, and comparing males and females, we have found that males grow significantly faster than females, and that length-at-age for both sexes exhibits extreme length observations. Comparisons indicated that a Student-t model with mixed effects describes best back-calculated data regarding pink cusk-eel. This framework merged the strengths of different approaches to estimate growth parameters in harvested fish populations, considering modeling of individual variability of length-at-age, Bayesian inference, and skew distribution of errors from the Student-t model.

Wednesday 29.08.2018 11:00 - 12:30 Parallel Session F – COMPSTAT2018

CI012 Room Cuza Hall ROBUST STATISTICAL METHODS Chair: Alfio Marazzi

C0324: **Robust sparse maximum association estimators**
Presenter: **Andreas Alfons**, Erasmus University Rotterdam, Netherlands
Co-authors: Christophe Croux, Peter Filzmoser

The maximum association between two multivariate random variables is defined as the maximal value that a bivariate association measure between respective one-dimensional projections of each random variable can attain. Using the Spearman or Kendall rank correlation as projection index thereby yields a more robust procedure than using the Pearson correlation. We propose a projection pursuit algorithm based on alternating series of grid searches in two-dimensional subspaces of each data set, together with an extension that allows for sparse estimation of the projection directions to increase the interpretability of the results in higher dimensions. In addition, we provide a fast implementation of the algorithm for the statistical computing environment R.

C0331: **Detection of outbreaks in notifiable disease data**
Presenter: **Matias Salibian-Barrera**, The University of British Columbia, Canada
Co-authors: Tae Yoon Lee

In order to detect and control possible disease outbreaks, the British Columbia Centre for Disease Control (BC CDC) monitors approximately 60 reportable disease counts from 8 branch offices in 16 health service delivery areas. Instead of relying on the judgement of staff on whether the reported numbers are higher than expected, in the early 2000s the BC CDC commissioned an automated statistical method to detect outbreaks in notifiable disease counts. To accommodate the seasonality of some diseases like meningococcal infections, longer term trends such as the periodicity of pertussis cases, and the decline in diseases like acute hepatitis B, this method is based on a generalized partially linear additive model. However, the current approach relies on certain ad-hoc criteria, and the BC CDC is interested in considering other alternatives. We discuss an outbreak detection method based on robust estimators of the distribution of the number of cases of each disease. The proposal builds on recently proposed robust estimators for additive, generalized additive, and generalized linear models. Using real and simulated data we compare our method with the approach currently used by the BC CDC and other natural competitors.

C0326: **Robust sparse principal components analysis**
Presenter: **Stefan Van Aelst**, University of Leuven, Belgium
Co-authors: Yixin Wang

Sparse principal component analysis can be used to obtain stable and interpretable principal components from high-dimensional data. Robust sparse PCA is considered to handle outliers in the data. The new method LTS-SPCA starts from the MLTS-PCA method which provides a robust but non-sparse PCA solution. MLTS-PCA yields the PC subspace corresponding to the proportion of the data which gives the smallest sum of squared residuals. To get sparse solutions, LTS-SPCA then incorporates an l1-norm penalty on the loading vectors to obtain sparsity. LTS-SPCA searches for the PC directions sequentially. This approach avoids that score outliers in the PC subspace destroy the sparse structure of the loadings. Simulation studies and real data examples show that LTS-SPCA can give accurate estimates, even when the data is highly contaminated. Moreover, compared to existing robust sparse PCA methods, LTS-SPCA can reduce the computation time to a great extent.

CO054 Room Clio Hall RECENT ADVANCES IN MIXTURE MODELING AND MISSING DATA ANALYSIS Chair: Xinyuan Song

C0225: **Bayesian two-part model for semicontinuous data with latent variables**
Presenter: **Xiaoqing Wang**, The Chinese University of Hong Kong, Hong Kong
Co-authors: Xinyuan Song

A joint modeling approach is proposed to investigate the observed and latent risk factors of semicontinuous responses of interest. The proposed model consists of two major components. The first component is a structural equation model (SEM), which characterizes latent variables through multiple observed variables and simultaneously assesses interrelationships among the latent variables. The second component is a two-part model for investigating the effects of observable and latent variables on semicontinuous responses of interest. The two-part model comprises a model for a binary indicator variable and a model for another response variable that is conditioned on the binary indicator variable. A full Bayesian approach with Markov chain Monte Carlo algorithm is developed for statistical inference. A simulation study demonstrates the satisfactory performance of the developed methodology. The method is then applied to a study concerning the relationship among non-cognitive abilities, education, and annual income.

C0266: **Cox regression with Potts-driven latent clusters**
Presenter: **Alejandro Murua**, University of Montreal, Canada
Co-authors: Danae Martinez-Vargas

A Bayesian nonparametric survival regression model with latent partitions is considered. The goal is to predict survival, and to cluster survival patients within the context of building prognosis systems. We propose the Potts clustering model as a prior on the covariates space so as to drive cluster formation on individuals and/or Tumor-Node-Metastasis stage system patient blocks. For any given partition, our model assumes an interval-wise Weibull distribution for the baseline hazard rate. The number of intervals is unknown. It is estimated with a lasso-type penalty given by a sequential double exponential prior. Estimation and inference are done with the aid of MCMC. To simplify the computations, we use the Laplace's approximation method to estimate some constants, and to propose parameter updates within MCMC. We illustrate the methodology with an application to cancer survival.

C0156: **Scalar on image regression with nonignorable missing data**
Presenter: **Xinyuan Song**, Chinese University of Hong Kong, Hong Kong

A scalar-on-image regression model is considered that uses ultrahigh dimensional imaging data as explanatory covariates. The model is used to investigate important risk factors for the scalar response of interest, which is subject to non-ignorable missingness. We propose the use of an efficient functional principal component analysis method to reduce the dimensions of the imaging observations. Given that non-ignorable non-response distorts the accuracy of statistical inference and generates misleading results, we propose an imaging exponential tilting model for the examination of the potential influence of imaging observations along with scalar variables on the probability of missingness. An instrumental variable, such as a covariate associated with the response but conditionally independent of the probability of missingness, is introduced to facilitate model identifiability. Statistical inference is conducted in a Bayesian framework with Markov chain Monte Carlo algorithms. Simulation studies show that the proposed method exhibits satisfactory finite sample performance. The methodology is applied to a study on the Alzheimer's Disease Neuroimaging Initiative dataset.

C0268: **Ampliclust: A fully probabilistic model-based approach denoising Illumina amplicon data**
Presenter: **Karin Dorman**, Iowa State University, United States
Co-authors: Xiyu Peng

Next-generation amplicon sequencing is a powerful tool for understanding microbial communities. Downstream analysis is often based on the construction of Operational Taxonomic Units (OTUs) with dissimilarity threshold 3%. The arbitrary threshold and reliance on OTU references can lead to low resolution, false positives, and misestimation of microbial diversity. We introduce Ampliclust, a reference-free method to resolve the number, abundance and identity of distinct variants sequenced in Illumina amplicon data. Unlike existing methods, Ampliclust is a fully probabilistic model, allowing the data to drive the conclusions rather than an algorithm or an external database. We use a modified Bayesian information criterion to estimate the number of sequence variants, and obtain maximum likelihood estimates of the abundance and identity of variants. Our model is able to match the performance of existing methods on well-separated mock communities, but achieves better accuracy in simulated communities with more similar variants. The major challenge for using mixture models in this context is the computational scalability to datasets consisting of millions or billions of observations in tens to thousands of clusters, which we begin to address through principled iterative schemes and improved initialization methods.

CO062 **Room Mezzanine Lounge** LARGE DATA SETS: METHODOLOGY AND APPLICATIONS **Chair: Malgorzata Bogdan**

C0323: **Deep Bayesian regression**

Presenter: **Florian Frommlet**, Medical University Vienna, Austria

Co-authors: Aliaksandr Hubin, Geir Olve Storvik

One of the most exciting recent developments in data analysis is deep learning. Multilayer networks have become extremely successful in performing prediction tasks and are successfully applied in many areas. However, the resulting prediction models often difficult to interpret and potentially suffer from overfitting. We bring the ideas of deep learning into a statistical framework which yields more parsimonious models and allows us to quantify model uncertainty. To this end we introduce the class of deep Bayesian regression models (DBRM) consisting of a generalized linear model combined with a comprehensive non-linear feature space, where non-linear features are generated just like in deep learning. DBRM can easily be extended to include latent Gaussian variables to model complex correlation structures between observations, which seems to be not easily possible with existing deep learning approaches. Two different algorithms based on MCMC are introduced to fit DBRM and to perform Bayesian inference. The predictive performance of these algorithms is compared with a numerous state of the art learning algorithms. Furthermore, we illustrate how DBRM can be used for model inference in various applications.

C0328: **Generalized information criterion in high-dimensional model selection**

Presenter: **Wojciech Rejchel**, University of Warsaw, Poland

Co-authors: Piotr Pokarowski, Agnieszka Prochenka, Michal Frej, Jan Mielniczuk

Model selection is a fundamental challenge for data sets that contains (much) more predictors than the sample size. In many practical problems (from genetics or biology) finding a (small) set of significant predictors is as important (or even more) as accurate estimation or prediction. The screening-selection algorithm is presented that is based on minimization of the empirical risk with the lasso penalty in the first step and with the generalized information criterion in the second step. We prove model selection consistency of this procedure in a wide class of models containing generalized linear models, quantile regression and support vector machines. The quality of the procedure is also investigated in numerical experiments.

C0348: **A power analysis for Knockoffs with lasso statistics**

Presenter: **Asaf Weinstein**, Stanford University, United States

Co-authors: Rina Foygel Barber, Emmanuel Candes

Knockoffs is a new framework for controlling the false discovery rate (FDR) in multiple hypothesis testing problems involving complex statistical models. While rigorous results have been obtained regarding type-I error control in a wide range of models, type-II error rates have been far less studied. In general, power calculations are admittedly difficult, in part owing to the very particular structure of the knockoff matrix. Nevertheless, there is a specific setting, involving an i.i.d. Gaussian design, where such calculations are possible. Working in that setting, we leverage recent results to show that a knockoff procedure associated with the Lasso path, achieves close to optimal power with respect to an appropriately defined oracle. This result demonstrates that, in our setting, augmenting the design with fake (knockoff) variables does not have a high cost in terms of power.

C0415: **Weeding out early false discoveries along the lasso path via knockoffs**

Presenter: **Malgorzata Bogdan**, University of Wroclaw, Poland

Co-authors: Weijie Su, Asaf Weinstein, Emmanuel Candes

LASSO is one of the most popular methods for identifying predictors in large data bases. This happens despite the fact that in practical applications LASSO often returns many false discoveries (i.e. variables which in fact are not correlated with the response). Recently this fact has been theoretically described using the framework of linear sparsity regime of Approximate Message Passing Theory for Gaussian designs. Specifically, a precise trade-off between the power and the false discovery rate has been provided, which holds independently of the signal magnitude and can not be broken for any value of the tuning parameter. We will show that this limitation can be removed by thresholding the solution of LASSO. The appropriate threshold can be obtained using the knock-off methodology, which allows us to control the False Discovery Rate. We will present theoretical results showing that this approach allows us to break through the FDR-Power Diagram for any given value of the tuning parameter. We will also empirically demonstrate that selecting the tuning parameter by cross-validation allows us to obtain a roughly optimal power for a given FDR level.

CO026 **Room Vega Hall** MULTIVARIATE ANALYSIS AND BIG DATA **Chair: Dominique Guegan**

C0202: **News, volatility and price jumps**

Presenter: **Massimiliano Caporin**, University of Padova, Italy

Co-authors: Francesco Poli

From two professional news providers we retrieve news stories and earnings announcements of the SP100 constituents and 10 macro fundamentals, moreover we gather Google Trends of the assets. We create an extensive and innovative database, useful to analyze the link between news and asset price dynamics. First, we shed light on the impact of information measures on daily realized volatility and select them by penalized regression. We also use them to forecast volatility and obtain superior results with respect to models that omit them. Second, we fit penalized logistic regression linking the probability of intraday jumps occurrence to news indicators. Announcements and surprises associated with FOMC rate decisions, federal budget, natural gas stocks and ECRI are potential causes of jumps. Further, news stories, i.e. not macro-related, are also relevant, in particular M&A and earning announcements.

C0206: **On the parameters estimation of the new Seasonal FISSAR model and simulations**

Presenter: **Papa Ousmane Cisse**, LMM Le Mans University / Paris 1 University, Senegal

Co-authors: Dominique Guegan, Abdou Ka Diongue

The model called Fractionally Integrated Separable Spatial Autoregressive processes with Seasonality, and denoted Seasonal FISSAR, for two-dimensional spatial data is considered. This new model will be able to take into account long memory in spatial data and periodic or cyclical behaviours presented in a lot of applications, including temperatures, agricultural data, epidemiology when the data are collected during different seasons at different locations, and also financial data, to take into account the specific systemic risk observed on the global market. We discuss on the methods of estimating the parameters of the Seasonal FISSAR model and show the asymptotic properties. First, we implement the regression method based on the log-periodogram and the classical Whittle method for estimating memory parameters. For estimating all model parameters simultaneously, innovation parameters and memory parameters MLE and Whittle method based on the MCMC are considered. We show the consistency empirically, and we investigate the asymptotic normality of the estimators by simulations. The first motivation behind MLE is to estimate all parameters simultaneously unlike our proposal regression method and Wittle method. However, the computational complexity of MLE may be outweighed by its convenience as a very widely applicable method of estimation.

C0215: **Credit risk analysis using machine and deep learning models**

Presenter: **Dominique Guegan**, Universite Paris 1 - Pantheon-Sorbonne, France

Due to the advanced technology associated with Big Data, data availability and computing power, most banks or lending institutions are renewing their business models. Credit risk predictions, monitoring, model reliability and effective loan processing are key to decision-making and transparency. We build binary classifiers based on machine and deep learning models on real data in predicting loan default probability. The top 10 important features from these models are selected and then used in the modelling process to test the stability of binary classifiers by comparing their performance on separate data. We observe that the tree-based models are more stable than the models based on multilayer artificial neural networks. This opens several questions relative to the intensive use of deep learning systems in the enterprises.

C0368: **Sparse nonparametric dynamic graphical models**

Presenter: **Fabrizio Poggioni**, University La Sapienza, Italy

Co-authors: Mauro Bernardi, Lea Petrella

A Sparse Nonparametric Dynamic Graphical Model is proposed for financial applications in which we employ a semiparametric multiple quantile model with CAViaR specification to describe the marginal distributions and a LASSO-penalized Gaussian Copula-VAR model to describe the multivariate distribution of financial returns as a sparse dynamic model. In order to use the multiple quantile models as marginal distributions the estimated quantile functions must be invertible, in this way we can get the marginal CDFs from the estimated multiple quantiles. It is therefore necessary to guarantee the monotonicity of the estimated quantiles and consequently the absence of crossing. Hence, we offer a contribution to the theme of quantile crossing for semiparametric models by defining a non-crossing parametric space for multiple quantile CAViaR models. Furthermore, we find computationally convenient to include the defined non-crossing necessary conditions as linear constraints to the multiple quantile estimation problem. Finally, we present an empirical application of the proposed methodology.

CG011 Room Orion Hall DEVELOPMENTS FOR DISCRETE DATA — Chair: Sebastian Doehler

C0165: **Robust estimation of a dynamic spatiotemporal model with count data**

Presenter: **Charlene Mae Celoso**, University of the Philippines Diliman, Philippines

A dynamic spatial-temporal model with a count dependent variable is estimated with a hybrid of forward search algorithm and bootstrap embedded into the backfitting algorithm. The method is evaluated for its robustness in some count data generating process. Simulation studies indicated that the method perform better in terms of MAPE when there are more time points than observations units in space and when the covariates contributes more than the spatial externalities.

C0176: **Estimating a Poisson autoregressive model with the backfitting algorithm**

Presenter: **Paolo Victor Redondo**, University of the Philippines Diliman, Philippines

Co-authors: Erniel Barrios, Joseph Ryan Lansangan

A Poisson autoregressive model (PAR) that accounts for discreteness and autocorrelation of count time series data is typically estimated within the context of state-space modelling with maximum likelihood estimation (MLE). The complexity of dependencies exhibited by count time series data however, complicates MLE. PAR is viewed as an additive model and is estimated using a hybrid of conditional least squares and MLE in the backfitting framework. Simulation studies show that estimation of PAR model viewed as an additive model is always better than PAR model in the state-space context whenever the non-normality of covariates for the latter is evident. In cases where the MLE of the PAR model in the state-space context exists, the estimates are comparable with the proposed method. The proposed method is then used in modelling incidence of tuberculosis, elucidating the role of various stakeholders in curbing the prevalence rate of the disease.

C0180: **Nonparametric test for intervention effect on count data**

Presenter: **Marcus Jude San Pedro**, School Statistics, University of the Philippines Diliman, Philippines

Co-authors: Erniel Barrios, Joseph Ryan Lansangan

Intervention effect triggers possible alteration in the behavior of the system, this is particularly highlighted in count time series data. We postulate an integer-valued autoregressive process and incorporates a temporary structural change as additive intervention effect. We propose a nonparametric test for the significance of intervention effect via the empirical distribution of test statistic through sieve bootstrap. Simulation study shows the nonparametric test exhibits high power and correct size especially when the intervention effect does not persist through time.

C0188: **Controlling the false discovery rate for discrete data: New results and computational tools**

Presenter: **Sebastian Doehler**, Darmstadt University of Applied Science, Germany

Co-authors: Etienne Roquain, Guillermo Durand

The Benjamini-Hochberg procedure and related methods are classical methods for controlling the false discovery rate for multiple testing problems. These procedures were originally designed for continuous test statistics. However, in many applications, the test statistics are discretely distributed. While it is well known that e.g. the Benjamini-Hochberg procedure still controls the false discovery rate in the discrete paradigm, it may be unnecessarily conservative. Thus, developing more powerful FDR procedures for discrete data is interesting. We present improved procedures that incorporate the discreteness of the p-value distributions and introduce an R package which implements these approaches.

CP117 Room Cocktail Hall POSTER SESSION II — Chair: Panagiotis Paoullis

C0178: **A family of the adjusted estimators maximizing the asymptotic predictive expected log-likelihood**

Presenter: **Haruhiko Ogasawara**, Otaru University of Commerce, Japan

A family of the estimators adjusting the maximum likelihood estimator by a higher-order term maximizing the asymptotic predictive expected log-likelihood is introduced under possible model misspecification. The negative predictive expected log-likelihood is seen as the Kullback-Leibler distance plus a constant between the adjusted estimator and the population counterpart. The vector of coefficients in the correction term for the adjusted estimator is given explicitly by maximizing a quadratic form. Examples using typical distributions in statistics are shown.

C0265: **Reliability analysis of switching parameters in resistive random access memories**

Presenter: **Ana Maria Aguilera**, University of Granada, Spain

Co-authors: Christian Acal, Juan Eloy Ruiz-Castro, Francisco Jimenez-Molinos, Juan Bautista Roldan

In the last few years, resistive random access memories (RRAMs) have been proposed as one of the most promising candidates to overcome the current Flash technology in the market of non-volatile memories, because of having optimal properties and outstanding possibilities for fabrication in the present CMOS technology. The stochastic nature of the physical processes behind the operation of resistive memories makes variability one of the key issues to solve from the industrial viewpoint of these new devices. Nowadays and with the purpose of extracting information about this issue, the electronic industry makes use of the Weibull distribution to model reset and set voltages from experimental data measured in these devices. However, in many occasions the weibull fit is not enough accurate and therefore, a different probability distribution must be considered. A new statistical approach based on phase-type modelling is proposed to get better fit and parameter interpretation.

C0270: **High-dimensional GMANOVA test**

Presenter: **Takayuki Yamada**, Kagoshima university, Japan

Co-authors: Tetsuto Himeno

This study treats the problem for testing generalized multivariate linear hypothesis for location parameter matrix. We proposed a test which is applicable for the case in which the dimension of the observed matrix is greater than the sample size. The limiting null distribution of the test statistic is derived and the power of the test is studied. The simulation results show that the test significantly outperforms well. Analysis of a rat weight data is carried out to demonstrate the proposed testing procedure.

C0362: **Website and text classification using natural language processing**

Presenter: **Martin Wood**, Office for National Statistics, United Kingdom

Company websites provide copious information on the economic activity of businesses, useful when measuring the modern economy. This information is in the form of natural language, making automated harvesting difficult. We review our experience with different statistical models generated to represent the documents (web pages) we wish to classify, and the methods and algorithms we can then apply to these representations. It is necessary to find a representation which extracts information at the level of abstraction of the concept we wish to classify or analyse within the document. For example, in determining if a website is used for sales, Bag-Of-Word (BOW) and Term Frequency-Inverse Document Frequency (TF-IDF) models allow classifiers to identify diagnostic keywords. Conversely, if we wish to identify more general business activity (e.g., mining), sophisticated representations built using Latent Dirichlet Allocation (LDA) or word2vec that characterise word associations are required. These representations can be extended with part-of-speech tagging, n-grams, metadata about the location of occurrence of the text on the website structure and by combining representations together in to larger feature sets. In all cases, sophisticated classifiers such as Random Forests and Support Vector Machines that are robust in the face of high-dimensional data perform best in supervised classification of the data.

C0372: **Kronecker-type products useful in constructions of non-orthogonal split-plot x split-block designs**

Presenter: **Iwona Mejza**, Poznan University of Life Sciences, Poland

Co-authors: Katarzyna Ambrozy-Deregowska

The split-plot x split-block design is a known design in field experimentation that is often used to carry out three-factor experiments. It is called a mixed SPSB design, and in the complete (orthogonal) version it is a combination of split-plot and split-block orthogonal designs. Often, however, due among other things to limitations of the experimental material, experiments are planned in incomplete (non-orthogonal) designs. To obtain an incomplete SPSB design with optimal efficiency for selected contrast groups, different construction methods and other experimental designs, so-called generating designs with known statistical properties, belonging to the class of block designs, are used. The incompleteness of a generated design planned in advance may relate to one factor, two factors or all three. To compare methods, an ordinary Kronecker product and a semi-Kronecker (Khatri-Rao) product were used in construction procedures. With both methods, some generated SPSB designs were obtained where the levels of each factor were allocated to a balanced square lattice design. A comparison of the methods of constructing designs, with respect to stratum efficiency factors for some contrasts of treatment parameters and with respect to the size of an experiment, is given.

C0391: **The multivariate statistic based on combining the t test and Wilcoxon test**

Presenter: **Masato Kitani**, Tokyo University of Science, Japan

Co-authors: Hidetoshi Murakami

A multivariate two-sample testing problem is one of the most important topics in nonparametric statistics. The multivariate nonparametric tests based on the Jureckova-Kalina ranks of distance have been discussed. For univariate case, it is proposed that a maximum test combining the t test and Wilcoxon's rank-sum test. It was shown that the maximum test has a good power for a variety of distributions, and its power is close to that of more powerful of the two tests. We obtain the null distribution of maximum test and extend the maximum test for the multivariate observation by using the Jureckova-Kalina ranks of distances. Simulations are used to investigate the power of suggested test for the two-sided alternative with various distributions. The behavior of proposed test is compared with the Hotelling's T^2 test. The results show that the proposed test statistic is more suitable than various existing statistics for testing a shift in the location and location-scale parameters.

C0392: **A new scale estimator based on the kernel density estimator with ranked set sampling**

Presenter: **Hikaru Yamaguchi**, Tokyo University of Science, Japan

Co-authors: Hidetoshi Murakami

The ranked set sampling (RSS) is a cost-efficient alternative to the simple random sampling (SRS) in situations where the exact measurements of sample units are difficult or expensive to obtain but (judgment) ranking of them according to the variable of interest is relatively easy and cheap. We investigate the point estimation and interval estimation of scale by using RSS data for two independent random variables with unknown probability distributions. Kernel density estimation is one of nonparametric probability density estimation methods and is used in various fields. In the literature, the estimation of the probability $P(X<Y)$ for the location parameter based on the kernel density estimator is proposed and its analog in RSS is also discussed. We propose a new scale estimator based on the kernel density estimation with RSS and show that the suggested estimator is superior to that of SRS. In addition, we show the consistency and the asymptotic unbiasedness of the proposed estimator. In point estimation, the performance of estimators is compared by mean square error. A bootstrap method based on RSS is used to construct the nonparametric confidence intervals and the performance of the proposed estimator is evaluated by Monte Carlo simulation.

C0393: **The multivariate rank tests based on the likelihood ratio**

Presenter: **Soshi Kawada**, Tokyo University of Science, Japan

Co-authors: Hidetoshi Murakami

Testing hypothesis is one of the most important topics in nonparametric statistics. Various nonparametric tests have been proposed for two-sample and multisample testing problems involving location, scale and location-scale parameters. It is well known that the nonparametric rank statistics based on the likelihood ratio is powerful and robust statistics. Recent progress in computerized measurement technology has permitted the accumulation of multivariate data, increasing the importance of multivariate data in many scientific fields. Thus, a multivariate examination of the data is very appropriate. However, in many applications, the underlying distribution cannot be assumed to follow a specific distribution, and nonparametric hypothesis testing must be used. For multivariate data, it is important to determine how to represent rank based on observation distances. We apply the extended Jureckova-Kalinas rank of distance to the rank statistics based on the likelihood ratio. Jureckova-Kalina rank of distance is invariant under affine transformation with respect to the shifted location. We compare the powers of the proposed test with the multivariate two-sample and multisample nonparametric tests for various distributions by using simulation studies.

C0408: **An extended block restricted isometry property for sparse recovery with non-Gaussian noise**

Presenter: **Klara Leffler**, Umea University, Sweden

Co-authors: Zhiyong Zhou, Jun Yu

Recovering an unknown signal from significantly fewer measurements is a fundamental aspect in computational sciences today. The key ingredient is the sparsity of the unknown signal, a realisation that that has led to the theory of compressed censing, through which successful recovery of high dimensional (approximately) sparse signals is now possible at a rate significantly lower than the Nyquist sampling rate. Today, an interesting challenge lies in customizing the recovery process to take into account prior knowledge about e.g. signal structure and properties of present noise. We study recovery conditions for block sparse signal reconstruction from compressed measurements when partial support information is available via weighted mixed l2/lp minimization. We show theoretically that the extended block restricted isometry property can ensure robust recovery when the data fidelity constraint is expressed in terms of an lq norm of the residual error. Thereby, we also establish a setting wherein we are not restricted to a Gaussian measurement noise. The results are illustrated with a series of numerical experiments.

C0430: **A maximum likelihood estimation of Brun's constant under the twin primes distribution**

Presenter: **Ryuichi Sawae**, Okayama University of Science, Japan

Co-authors: Daisuke Ishii

The results of research on the distribution of prime numbers are numerous, including a very important theorem in number theory. On the other hand, twin primes are pairs of primes of the form $(p, p+2)$, but it has not been proven that there are infinite number of twins primes until now. Let the set K_2 be $\{(3,5),(5,7),(11,13),(17,19),\dots\}$ of twin-prime pairs. Then, although the sum of the reciprocals of all the primes is divergent, the sum of the reciprocals of K_2, i.e. $B_2 = (1/3+1/5)+(1/5+1/7)+(1/11+1/13)+(1/17+1/19)+\dots$, is convergent. The limit of this sum is known as Brun's sum or Brun's constant. For the maximum likelihood estimation of Brun's constant, we list up the twin primes by a fast Eratosthenes sieve, we use a fast computer algorithm for the reciprocals calculation and the error analysis for the Hardy-Littlewood approximation.

C0432: **Conditional skewness and kurtosis in the good and bad times**

Presenter: **Lukas Fryd**, University of Economics, Prague, Czech Republic

The main contribution is to the broader understanding of risk premium in the PX50 index before, during and after the crisis in 2008. The most common framework in financial economics take the variance as the main source of risk, and so most of the papers contain the models from the GARCH family. The GARCH helps to reduce the leptokurtosis of standardized returns but not eliminate it. We could find additional information in the high–order moments. If we assume that higher moments represents the additional risk premium, then we could get a better understanding of financial markets. We will assume the data generation process of returns follows the skewed generalized t distribution with time-varying parameters. The GARCH methodology will be utilized to the modeling of high-order moments as autoregressive processes.

C0443: **Empirical comparison of some automatic ARIMA modeling**

Presenter: **Dedi Rosadi**, Universitas Gadjah Mada, Indonesia

In some application of time series modeling, it is necessary to obtain forecast of various types of data automatically and possibly, in real-time way, for instance, to do a real-time processing of the satellite data. Various automatic algorithms for modeling ARIMA models are available in the literature, where we will discuss four methods in particular. One of the methods is based on a combination between the best exponential smoothing models to obtain the forecast, together with state-space approach of the underlying model to obtain the prediction interval. Other method, which is more advanced method, is based on X-13-ARIMA-SEATS, the seasonal adjustment software by the US Census Bureau. Two other methods use more heuristic approaches, namely genetic algorithms and the neural networks. These approaches are implemented in our R-GUI package, namely RcmdrPlugin.SPSS. We provide empirical application of the methods and tool using real data.

Wednesday 29.08.2018 14:15 - 16:15 Parallel Session G – COMPSTAT2018

CI113 Room Cuza Hall NEW COMPUTATIONAL METHODS FOR STATISTICAL INFERENCE Chair: Jaeyong Lee

C0445: **Fusion learning and confederate inference**

Presenter: **Peter Song**, University of Michigan, United States

Data analytics and statistical algorithms in data integration are considered. As data sets of related studies become more easily accessible, combining data sets of similar studies is undertaken in practice to achieve Big Data and to enjoy more powerful analysis. A major challenge arising from integrated data analytics pertains to principles of information aggregation, learning data heterogeneity, algorithms for model fusion. Information aggregation has been studied extensively by many statistics pioneers, which lay down the foundation of data integration. Also, ignoring such heterogeneity in data analysis may result in biased estimation and misleading inference. Distributed computing and inference will be discussed.

C0448: **Uncertainty quantification of treatment regime in precision medicine by confidence distributions**

Presenter: **Min-ge Xie**, Rutgers University, United States

Personalized decision rule in precision medicine can be viewed as a discrete parameter, for which theoretical development for statistical inference is lagged behind. A new way to quantify the estimation uncertainty in a personalized decision using confidence distribution (CD) is proposed. Specifically, in a regression setup, suppose the decision for treatment vs control for an individual x_a is determined by a linear decision rule $D_a = I(x_a b > x_a d)$, where b and d are unknown regression coefficients in models for potential outcomes of treatment and control, respectively. The data-driven decision $\hat{D}_a$ relies on the estimates of b and d and has uncertainty. We propose to find a CD for $c_a = x_a b - x_a d$ and compute a confidence measure of $\{D_a = 1\} = \{c_a > 0\}$. This measure is in $(0,1)$ and provides a frequency-based assessment on how reliable our decision is. For example, if the confidence measure of the decision $D_a = 1$ is 63%, then we know, out of 100 patients who are the same as patient x_a, 63 will benefit to have the treatment and 37 will be better off to be in the control group. Numerical study suggests that this new measurement is inline with classical assessments (i.e., sensitivity, specificity), but different from the classical assessments, this measurement can be directly computed from the observed data. Utility of this new measure will also be illustrated in an application of an adaptive clinical trial.

C0449: **Tracking functional data by nonparametric tolerance tubes**

Presenter: **Regina Liu**, Rutgers University, United States

Tolerance intervals and tolerance regions are important tools for process monitoring or statistical quality control of univariate and multivariate data, respectively. We discuss their generalization to tolerance tubes in the infinite dimensional setting for functional data. In addition to the generalizations of the commonly accepted definitions of the tolerance level of beta-content or beta-expectation, we introduce the new definition of alpha-exempt beta-expectation tolerance tube. The latter loosens the definition of beta-expectation tolerance tube by allowing alpha (pre-set using domain knowledge) portion of each functional be exempt from the requirement. Those proposed tolerance tubes are completely nonparametric and broadly applicable. We discuss their general properties, and show that the alpha exempt beta-expectation tolerance tube is particularly useful in the setting where occasional short term aberrations of the functional data are deemed acceptable (or unpreventable) and they do not cause substantive deviation of the norm. This desirable property is elaborated further and illustrated with both simulations and real applications in continuous monitoring of blood glucose level in diabetes patients as well as of aviation risk patterns of aircraft landings. This nonparametric tolerance tube can apply broadly to many domains dealing with functional data.

CO098 Room Cocktail Hall OPTIMAL EXPERIMENTAL DESIGN AND APPLICATIONS Chair: Frank Miller

C0306: **Optimal designs for complex problems**

Presenter: **Weng Kee Wong**, UCLA, United States

Algorithms are practical ways to find optimal experimental designs. Most published work in the statistical literature concern optimal design problems for a model with few factors or assume the model is additive when there are several factors. Nature-inspired metaheuristic algorithms are general and powerful optimization tools that seem to be under-utilized in statistical research. We describe a few of these algorithms, list their advantages over current methods and present optimal designs for a few complex biostatistical problems with real world applications.

C0226: **Optimal designs for dose response curves with common parameters**

Presenter: **Kirsten Schorning**, Ruhr-University Bochum, Faculty of Mathematics, Germany

Co-authors: Holger Dette, Bjoern Bornkamp, Chrystel Feller, Georgina Bermann

A common problem in Phase II clinical trials is the comparison of dose response curves corresponding to different treatment groups. If the effect of the dose level is described by parametric regression models and the treatments differ in the administration frequency (but not in the sort of drug) a reasonable assumption is that the regression models for the different treatments share common parameters. We develop optimal design theory for the comparison of different regression models with common parameters. We derive upper bounds on the number of support points of admissible designs, and explicit expressions for D-optimal designs for frequently used dose response models with a common location parameter. If the location and scale parameter in the different models coincide, the problem becomes much harder and therefore we determine minimally supported designs and sufficient conditions for their optimality in the class of all designs.

C0250: **Minimax optimum design for choosing between models for enzyme inhibition**

Presenter: **Ellinor Fackle-Fornius**, Department of Statistics, Sweden

In enzyme kinetics when studying inhibition, an extended version of the Michaelis-Menten model can be used to model two different types of enzyme inhibition: competitive and non-competitive inhibition. In order to discriminate between the two types of inhibition precise estimation of one of the model parameters is desired and a DS-optimum design would be suitable. Since the optimum design will be parameter dependent one option is to use the minimax principle for design optimality. The minimax design seeks the minimum of the maximum criterion function for a set of plausible parameter values specified by the researcher. Thus, it aims to be robust as long as the parameter values belong to the prespecified set. In this application, we derive minimax DS-optimum as well as maximin efficient designs (where design efficiency in relation to the locally optimal design is the criterion). We evaluate the minimax and maximin efficient designs in terms of variance and efficiency over the prespecified set of parameter values and make comparisons with locally optimum designs and another type of robust design.

C0211: **Optimal item calibration for computerized achievement tests**

Presenter: **Mahmood UL-Hassan**, Stockholm University, Sweden

Co-authors: Frank Miller

Item calibration is a technique to estimate characteristics of questions (called items) for achievement tests. In computerized adaptive tests (CAT), item calibration is an important tool for maintaining, updating and developing new items for an item bank. To efficiently sample examinees with specific ability levels for this calibration, we use optimal design theory where we assume that the probability to answer correctly to an item follows a two-parameter logistic model. A locally D-optimal unrestricted design for each item has two design points for ability. In practice, it is hard to sample examinees from a population with these specific ability levels due to unavailability or limited availability of examinees. To counter this problem, we use the concept of optimal restricted designs and show that this concept naturally fits to item calibration. Locally D-optimal restricted designs provide us two intervals of ability levels for optimal calibration of an item. Several scenarios with optimal restricted designs are presented here when one or two items have to be calibrated. These scenarios recommend us that the naive way to sample examinees around unrestricted design points is not the optimal way to calibrate an item.

C0357: **Optimal experimental design that minimizes the width of simultaneous confidence bands**
Presenter: **Satoshi Kuriki**, The Institute of Statistical Mathematics, Japan
Co-authors: Henry Wynn

An optimal experimental design is proposed for a curvilinear regression model that minimizes the band-width of simultaneous confidence bands. Simultaneous confidence bands for curvilinear regression are constructed by evaluating the volume of a tube about a curve that is defined as a trajectory of a regression basis vector. The proposed criterion is constructed based on the volume of a tube, and the corresponding optimal design that minimizes the volume of tube is referred to as the tube-volume optimal (TV-optimal) design. For Fourier and weighted polynomial regressions, the problem is formalized as one of minimization over the cone of Hankel positive definite matrices, and the criterion to minimize is expressed as an elliptic integral. We show that the Möbius group keeps our problem invariant, and hence, minimization can be conducted over cross-sections of orbits. We demonstrate that for the weighted polynomial regression and the Fourier regression with three bases, the tube-volume optimal design forms an orbit of the Möbius group containing D-optimal designs as representative elements.

CO056 Room Mezzanine Lounge **IFCS SESSION: METHODS FOR COMPLEX AND MIXED TYPE DATA** **Chair: Theodoros Chatzipantelis**

C0276: **On the performance of distance-based approaches for clustering mixed-type data**
Presenter: **Angelos Markos**, Democritus University Of Thrace, Greece
Co-authors: Odysseas Moschidis, George Menexes, Theodoros Chatzipantelis

Clustering of a set of objects described by a mixture of continuous and categorical variables is a challenging task. Popular distance-based approaches for clustering mixed type-data include dissimilarity measures for variables with different measurement scales, standardization of variables to the same scale, extensions of K-means for mixed data and sequential or simultaneous dimension reduction and clustering. A major concern in clustering of mixed-type data is how to achieve a favorable balance between continuous and categorical variables. A number of existing methods require user-specified weights to determine the relative contribution of continuous versus categorical variables. Other approaches adaptively adjust weights by considering the importance of each type of variables. We study the similarity of clustering solutions obtained by different strategies on a number of real mixed-type data sets and study their performance on simulated data sets with varying levels of continuous and categorical overlap. Dimension reduction and clustering methods tend to outperform alternative approaches when categorical variables are more informative than continuous for purposes of clustering, i.e. for data sets in which the continuous variables have substantially more overlap compared to the categorical ones. Recommended practices are provided within the context of this framework.

C0243: **A clustering method proposition for mixed type data**
Presenter: **Odysseas Moschidis**, University of Macedonia, Greece
Co-authors: Theodoros Chatzipantelis

The typical encoding of a continuous variable in a categorical ordinal variable, presents two major drawbacks: a) distinctively different values are classified in the same class with a major loss of information; b) values close to one another, which stand each side the boundary of two classes are classified in different classes, with a distortion of information. Few algorithms cluster mixed type datasets with both numerical and categorical attributes. We propose an algorithm that enables hierarchical clustering of data with numerical and categorical attributes based on WARD criterion and chi-squared metric. Each categorical variable is replaced by a set of 0-1 variables, one for each variable category, taking value 1 if the corresponding category has been observed and 0 otherwise and each numerical variables is replaced by a set of n-grades possibilities. With the proposed encoding that is an evolvement of ordinal data encoding, each value of the continuous variable could be classified in all n-classes of the categorical variable using as values the probabilities of a corresponding probability distribution function, different for each value of the numerical variable. This results in the elimination of the drawbacks a) and b) of the typical encoding, for, as we are going to suggest, we achieve the reconstruction of the values of the numerical variable. The proposed methodology gives similar results with MCA

C0426: **Missing data imputation problems may occur when dealing with categorical data**
Presenter: **Francesco Palumbo**, University of Naples Federico II, Italy
Co-authors: Alfonso Iodice D Enza, Angelos Markos

Missing data imputation issues may occur when dealing with categorical data: for example, in surveys, respondents are reluctant to answer questions related to sensitive information (e.g. income, sexual orientation, religion). When missing is related to some of the observed data and it only occurs in a subset of variables, missingness is referred in the literature as missing at random (MAR). Under these conditions, good techniques need to incorporate variables that are related to the missingness. In MAR imputation, several approaches work satisfactorily when dealing with continuous variables; however they cannot be easily generalized to multinomial categorical data. In this contribution a procedure is proposed that combines principal component methods and multiple imputation via chained equations (MICE) to impute missing entries in high dimensional categorical data. Given a set of p categorical variables and one variable with MAR values, the procedure imputes the missing entries according to association structure between the incomplete variable and the p observed variables. In particular, a reduced number of linear combinations (principal components) are defined by means of correspondence analysis-based methods; such components are the input of a suitable MICE procedure. The procedure can be iterated when more than one of the categorical variable presents missing entries. The contribution ends with a comparative study among other categorical data imputation approaches.

C0263: **A new criterion for forming compact groups in hierarchical clustering**
Presenter: **Theodoros Chatzipantelis**, Aristotle University Thessaloniki, Greece
Co-authors: Dimitrios Karapistolis

Analysing the political culture, our model is based on the axiomatic assumption that individual worldviews of the good and the evil, and of the sacred and the profane in political matters derives, but is not determined, from hegemonic cosmological and ontological principles which constitute the public sphere of meaning. These principles define a given political culture, though they do not need to be internally cohesive or comprehensive. Data analysis for a sample of 900 students was based on Hierarchical Cluster Analysis (HCA) and Multiple Correspondence Analysis (MCA). In the first step, HCA was used to assign subjects into distinct groups according to their response patterns. Furthermore, for each group, the contribution of each question (variable) to the group formation was investigated, in order to reveal a typology of behavioural patterns. To determine the number of clusters, two methods were used: FACOR and KARAP. The main differences between the two methods are the algorithm for the internal constellations and juxtapositions defined by subjects and variables. By method KARAP we argue that we succeed to form compact groups connecting each subject in each group to the level of the variable that contributes to groups formation.

C0245: **A new visualization of the results of the correspondence analysis with R programming**
Presenter: **Stratos Moschidis**, Hellenic Statistical Authority (ELSTAT), Greece
Co-authors: Athanasios Thanopoulos

It is well known that in the human and social sciences the characteristics studied are mainly nominal-categorical. It is exactly the field that the method of correspondence analysis is widely used. The correct reading of the results of the method by non-special users presents several times difficulties and interpretive misconceptions. The purpose is an absolutely clear interpretive visualization of the results of the CA method using the R language, so that the reading of the results becomes automatic without the combined use of the interpretative indicators COR-CTR and co-ordinates.

CO066 Room Orion Hall RESEARCH METRICS FOR INSTITUTIONAL PERFORMANCE EVALUATION Chair: Frederick Kin Hing Phoa

C0210: **An efficient analysis on the network coverage via dominating centrality sets**
Presenter: **Frederick Kin Hing Phoa**, Academia Sinica, Taiwan

Despite analyzing via any traditional measures on the network centrality, misleading suggestions might result when there exists more than one center in a network, which commonly happened nowadays. Sequential selection on the network centers usually failed to reveal the truly influential set of center nodes due to the curse of double counting. We introduced a new measure on the network coverage of the influential set of center nodes via a new graph-theoretic tool called the dominating centrality set (DCS). It simultaneously identified the center nodes and analyzed their coverage to the whole networks. Exact formulas on cycle counts from nodes were derived to provide computationally efficient calculations on the DCS. We demonstrated this method to the analysis of Amazon product network and the DBLP network, and hinted the potential use of this method towards the analysis of Web of Science database.

C0253: **A study of the article citation network in statistics research community**
Presenter: **LinHsuan Chang**, SOKENDAI (The graduate university for advanced studies), Taiwan
Co-authors: Junji Nakano, Frederick Kin Hing Phoa

Web of Science (WoS) database can be viewed as a big multi-layer, multi-level and dynamic network of articles over years. Among all layers of network, we are interested in some implicit phenomena in the article citation network in statistics research community. Our goal is to develop a method which can evaluate the research performance more accurately and can improve the problems of the current evaluation methods. It is clear that the citation network is described by directed graph with special structure according to the published time of articles. We use the theoretical approach to generate the network considering these structures explicitly. Then we introduce a new quantity to measure the importance or influence of an article, in terms of its multi-stage citation structure in the whole network. We undergo a systematic ranking process by using our method for all articles being tagged as statistics in the WoS subject area, identifying which articles are influential in statistics community within 2005-2014.

C0238: **Author name identification using a Dirichlet-multinomial regression topic model**
Presenter: **Tomokazu Fujino**, Fukuoka Women University, Japan
Co-authors: Keisuke Honda, Hiroka Hamada

A new framework is proposed for extracting a complete list of the articles written by researchers who belong to a specific research or educational institute from an academic document database such as the Web of Science. The framework is based on Latent Dirichlet Allocation (LDA), which is a topic model. To improve the framework we use various techniques and indices, such as synonym retrieval, inverse document frequency and Dirichlet-multinomial Regression (DMR). By using DMR, it is possible to reflect observed features of the articles such as author's affiliation in the topic distribution derived by LDA. A numerical example is presented to illustrate the framework, and we will discuss how much improvement is possible by using DMR topic model.

C0247: **A development of new metric on graph data using stochastic block model and pointwise mutual information**
Presenter: **Keisuke Honda**, The Institute of Statistical Mathematics, Japan
Co-authors: Hiroka Hamada, Mio Takei

A new clustering method is introduced to measure influence of papers in all areas of science and propose a new metric which has well useful properties such ad article level, field independent. To illustrate one application of our method, we analyzed over 7 million articles published between 2012 and 2016 from Web of Science(WOS). Our method consists two key techniques such as stochastic block model (SBM) and Pointwise mutual information (PMI). As first step, to see structure of entire relationship among papers we apply SBM on big scale citation network data. SBM generates a matrix which divides several blocks which represent relationship among research fields. This matrix can be defined as latent research subject based on completely co-citation network of academic activities in real-world. Secondary, to eliminate the influence of bias between research fields we apply PMI as normalization method. Finally, Diversity of papers is calculated by sum of value which correspond to elements in matrix as distance of latent and normalized research filed. The resulting our Research Diversity index (REDi) provides an alternative to the invalid evaluation of using journal impact factors to identify influential papers for various research filed and long-term effect.

C0235: **Understanding research trends based on paper abstracts using topic modeling**
Presenter: **Mio Takei**, The Institute of Statistical Mathematics, Japan
Co-authors: Tomokazu Fujino, Keisuke Honda, Junji Nakano

With the scale of real data becoming larger, the statistical science has become also important in the general society. We investigate research movements and trends in statistical science in the academic filed using Latent Dirichlet Allocation (LDA). Abstracts from academic papers, which is the most objective output of research activities, are available by Web of Science (WoS) and are good data to see trends of research. Therefore, we analyzed these data for understanding current research movements and trends in statistical science. We applied LDA model on paper abstracts and estimated topics. We then aggregated these topic distributions over time to find research movements. Our results can be used to understand the topics in statistical science that will be hot in the future (and also cold in the future). We can use these results to determining the special themes in our institute research to promote active statistical research in Japan. Furthermore, we can help researchers to discover promising research topics in the future.

CC081 Room Clio Hall APPLIED STATISTICS Chair: Agustin Mayo-Iscar

C0379: **Statistical methods to study consistency between declared and measured values on waste packages**
Presenter: **Luan Jaupi**, CNAM, France

Very precise technical and operational specifications are provided for the management of radioactive waste. For each manufactured waste package, the producer determines the activities of radionuclides that are present in the package, commonly called declared activities. During the storage process, compliance checks are performed on some packages and radioactivity measurements are made, called measured activities. Statistical methods are considered which are used to conclude consistency between declared and measured values on waste packages. To achieve this aim, we propose the use of different statistical methods: statistical bounds, regression analysis, hypothesis testing and analysis of recoveries.

C0230: **Costing system for nursing homes: A Portuguese case**
Presenter: **Ines Lopes**, ISCTE - IUL, Portugal
Co-authors: Abdul Suleman, Dalia Nogueira, Elizabeth Reis

A costing system for nursing homes is developed, based on the initial health evaluation of elderly people on admission. The data reflect the complexity items of the interRAI Long Term Care Facilities (LTCF) assessment form, and comprises $N = 387$ individuals institutionalized in six non-profit nursing homes in Portugal. A decomposing of the data in fuzzy clusters gives rise to a hierarchical fuzzy 3-partition: Low, Medium and High complexity. The particular distribution of individuals in this partition subsequently leads them to be indexed by a real number such that the higher the number the higher the complexity of the individual. Next, we solve the inverse problem by means of a linear regression model, which allows us to estimate the complexity of every individual on the basis of his/her initial health assessment. Then, we establish a relation between the current cost and the estimated complexity of individuals using another regression model. We add a second order term to account for the nonlinear effects, and obtain a model with good fit, $R^2_{adj} = 0.83$. We validate our model by considering 10 additional cases and find a good association between the estimated cost and the cost currently practiced, $p = 0.90$.

C0399: **Probabilistic forecasting of retail sales: A quantile regression approach**
Presenter: **Mikhail Zhelonkin**, Erasmus University Rotterdam, Netherlands

Accurate forecasting of sales is essential in retail for operations and management. The vast majority of research concentrates on point forecasts. In addition to point forecasting the forecasting of the entire distribution is helpful. It can be used for several purposes, including personnel scheduling, shelf replenishment, optimization of the shelf space, to mention a few. The data is characterized by strong seasonality, time varying volatility and skewness, presence of outliers (big orders) and natural contamination by out-of-stocks. Clear, that the use of robust methods in such circumstances is highly desirable. We propose to use the L_1 penalized quantile regression with harmonic functions as covariates. This allows us to solve two problems simultaneously: first, it provides the characterization of the distribution of sales, and second, under assumption of moderate number of out-of-stocks (less than 50%), we obtain a forecast of the actual demand.

C0378: **Dynamic prediction of propensity to purchase by landmark modelling**
Presenter: **Ilan Fridman Rojas**, Profusion Media Ltd, United Kingdom
Co-authors: Aris Perperoglou, Berthold Lausen, Henrik Nordmark

The aim is to present a novel application of a previous landmarking methodology to predict propensity to purchase based on transactional data. This use case presents a number of challenges, including data sets of considerable size for which many current statistical models and tools no longer scale to, and recurrent events with high frequency and multiplicity, often with time-varying covariates and strongly time-dependent effects. We present the results of such an application to subsets of a data set from a retailer with 2 million customers and 7 years of collected transactional data, extracting estimates of the time-varying effects, and producing dynamic predictions of probability of re-purchase which condition on time elapsed since the last purchase.

C0174: **Analysis of water-peptide dynamics using pair-copulas**
Presenter: **Jutharath Voraprateep**, Ramkhamhaeng University, Thailand

It is considered an established fact that water plays the major role in protein motion. There is a close connection between the water dynamics and the protein conformational dynamics. We report on statistical analysis of such conformational dynamics obtained using classical molecular dynamics simulations with explicit water. We investigate specific moments in time when one of the dihedral angles of a simulated protein makes a large amplitude change causing a conformational transition in the peptide. We are interested in finding statistical correlations between the values of the angle at the moment of transition and several moments in advance of the transition. We also investigate how these correlations change when conditioned on the presence of water at different locations in space around the peptide. The challenge is in a large number of parameters that influence the conformational dynamics, which leads to multivariate probabilities. As the statistical tools, we use pair-copulas and the Kendall's tau correlation. We have found that the dynamics of peptides conformation possesses temporal correlations well in advance of the moments of conformational transitions.

CC082 **Room Vega Hall** **TIME SERIES** **Chair: Alexander Duerre**

C0189: **A linear time method for the detection of point and collective anomalies**
Presenter: **Alexander Fisch**, Lancaster University, United Kingdom
Co-authors: Idris Eckley, Paul Fearnhead

Anomaly detection is of ever-increasing importance for many applications such as fault detection and fraud prevention. This is primarily due to abundance of sensors within contemporary systems and devices. Such sensors are capable of generating a large amount of data, necessitating computationally efficient methods for their analysis. To date, much of the statistical literature has been concerned with the detection of point anomalies, whilst the problem of detecting anomalous segments –often called collective anomalies– has been relatively neglected. We will introduce work that seeks to address this gap by introducing a linear time algorithm based on a parametric epidemic change point model. We present an approach that, with provable guarantees, is able to differentiate between both point anomalies and anomalous segments. Our computationally efficient approach is shown to be better than current methods, and we demonstrate its usefulness on the challenging problem of detecting exoplanets using data from the Kepler telescope.

C0349: **Recursive estimation of multivariate GARCH models**
Presenter: **Tomas Cipra**, Charles University, Prague, Czech Republic

Recursive algorithms for the parameter estimation and the volatility prediction in the multivariate GARCH models are suggested that seem to be useful for various financial time series, in particular for high frequency log returns which are (conditionally) correlated. These models are routinely estimated by computationally complex off-line estimation methods, e.g. by the conditional maximum likelihood procedure. However, in many empirical applications (especially in the context of UHF financial data) it seems necessary to apply on line methods which are numerically more effective to calibrate and monitor such models. Therefore, one (i) derives on-line estimation algorithms applying general recursive identification instruments for such models, and (ii) examines these methods by means of simulations and an empirical application.

C0429: **High-frequency volatility: Online and stream estimation perspective**
Presenter: **Vladimir Holy**, University of Economics, Prague, Czech Republic

An important aspect of financial high-frequency data analysis is volatility estimation. Volatility of the price process is typically measured by the quadratic variation (volatility realized over a given time period), integrated variance (quadratic variation of the continuous part of the process) and spot volatility (derivative of integrated variance). These quantities are typically estimated by non-parametric methods under rather general assumptions about the price process and in the presence of the market microstructure noise. A specific process can also be considered and parametric methods utilized. From the computational point of view, it is natural to consider high-frequency data to be a data stream. The question is which volatility estimators can be computed by online and stream algorithms.

C0438: **Generalized autoregressive score models for conditional quantiles**
Presenter: **Petra Tomanova**, University of Economics, Prague, Czech Republic

New specifications of conditional quantile models for value at risk are proposed. Our approach is based on Generalized Autoregressive Score (GAS) framework, also known as Dynamic Conditional Score (DCS) models, allowing parameters to vary over time and capturing the dynamics of time-varying parameters by the autoregressive term and the scaled score of the conditional observation density. Individual conditional quantiles are estimated directly without imposing distributional assumptions on returns. We compare our models to the originally proposed specifications of Conditional Autoregressive Value at Risk models. We pay a special attention to the estimation procedure. We compare various techniques how to initialize the estimation and address the issue of finding reasonable starting values since the optimization surface related to our problem is challenging. We also focus on assessing different link functions to derive models for conditional quantiles from GAS models, an approximation of quantiles of student t distribution so the estimation of conditional quantiles is computationally tractable, and other related computational issues.

C0416: **Spatial and spatiotemporal ARCH-type models**

Presenter: **Philipp Otto**, European University Viadrina, Germany

A general overview on spatial and spatiotemporal ARCH models is provided. To define spatial models, in particular areal spatial models like simultaneous autoregressive (SAR) models, it is convenient to consider a vector of observations $Y = (Y(s_1),\ldots,Y(s_n))'$ at all locations $s_1,\ldots,s_n$. For spatial ARCH models, we specify this vector as $Y = \text{diag}(h)^{1/2}\varepsilon$, analogue to the well-known time series ARCH models. However, note that the vector $h = (h(s_1),\ldots,h(s_n))'$ does not necessarily coincide with the conditional variance $Var(Y(s_i)|Y(s_1),\ldots,Y(s_{i-1}))$, as the variance in any location s_j for $j \neq i$ also depends on $Y(s_i)$. We now distinguish between several spatial ARCH-type models via the definition of h. In particular, we distinguish between three different spatial ARCH-type models. Beside the original definition, we introduce an exponential spatial ARCH model and propose maximum-likelihood estimators for the parameters of this new model. In addition, we consider a complex-valued definition of the spatial ARCH. From a practical point of view, the use of the R-package spGARCH is demonstrated.

Thursday 30.08.2018 09:00 - 11:00 Parallel Session H – COMPSTAT2018

CO036 Room Cocktail Hall CHALLENGES IN COMPUTATIONAL STATISTICAL MODELING AND RISK ASSESSMENT Chair: Ivette Gomes

C0166: **Bias reduction in tail estimation and modelling a full data set**

Presenter: **Jan Beirlant**, KULeuven, Belgium

In recent years several attempts have been made to model both the modal and tail part of the data. A dynamic mixture of two components with a weight function smoothly connecting the bulk and the tail of the distribution has been proposed. Recently, nice review on this topic has been made, and a new statistical model has been proposed which is in compliance with extreme value theory and allows for a smooth transition between the modal and tail part. Incorporating second order rates of convergence for distributions of peaks over thresholds, models have been constructed that can be viewed as special cases from both approaches discussed above. When fitting such second order models, it turns out that the bias of the resulting extreme value estimators is significantly reduced compared to the fit with one Pareto component. Recently, it has been shown that using penalized likelihood methods on the weight parameter one can obtain good bias and mean squared error properties for tail estimators. We encompass the above approaches providing models that can be used to model full data sets, that comply with extreme value theory, and that provide appropriate tail fits with special attention for tail estimation with reduced bias and mean squared error under the classical max-domain of attraction conditions.

C0236: **Exponential tilts for heteroscedastic extremes with an application to cryptocurrency markets**

Presenter: **Miguel de Carvalho**, FCiencias.ID (The University of Edinburgh / CEAUL, University of Lisbon), United Kingdom

Co-authors: Raphael Huser, Rodrigo Rubio

A density ratio model is proposed for modeling extreme values of non-identically distributed observations. The proposed model can be regarded as a proportional tails model for multisample settings. A semiparametric specification is devised so to link all elements in a family of scedasis densities through a tilt from a baseline scedasis. Inference is conducted by empirical likelihood inference methods. An application is given to model the dynamics of the frequency of extreme losses of some leading cryptocurrencies.

C0284: **Testing the suitability and sensitivity of parameters for a tumor morphology model**

Presenter: **Emma Turian**, Northeastern Illinois University, United States

Co-authors: Shuwang Li, John Lowengrub, Kai Liu, Kara Pham

The ability of tumors to metastasize is mostly preceded by morphological instabilities such as chains, or fingers of cancerous cells that invade the host environment. Therefore, parameters that control the morphological shape of the tumor contribute to its invasive ability. Earlier studies include mathematical modeling related to tumor dynamics. In order to describe the relationship between tumor and its host, tumor growth is analyzed using a two-phase Stokes model. Morphological changes are evaluated using two types of energy: The surface energy of the tumor-host interface, and the Helfrich bending energy, which allows the investigation of the stiffness of tumor-host interface. Using the bending energy approach, a modified Young-Laplace equation for the stress jump across the interface is developed through an energy variation approach. A linear stability analysis has been employed to assess the effects of viscosity, cell adhesion, bending rigidity, and apoptosis on tumor morphology. The model obtained suggests that increased tumor viscosity or apoptosis may lead to an unstable morphology. The suitability of the system parameters, as predicted by the model, has been analyzed using linear stability and sensitivity analysis. Comparison with experimental data found good agreement for certain categories of tumors.

C0336: **Tilting maximum Lq-estimation in the block maxima setting**

Presenter: **Claudia Neves**, University of Reading, United Kingdom

Over the last decade there has been an astonishing growth in the statistical techniques to analyse extreme values. As typified by the classical maximum likelihood (ML) inference on block maxima, the Generalised Extreme Value distribution is the appropriate probabilistic instrument when fitting a sample of maxima. The recent maximum Lq-likelihood method has a notable advantage to the usual ML estimation: with small up to moderate sample sizes, a proper choice for the distortion parameter $q > 0$ can deploy the variance in mitigating the mean squared error, thus eroding the role of bias. In this framework, an alternative class of parametric estimators stems from the maximum product of spacings (MSP) method through its obvious extension to the MSPq class. We will assess the current state of development and usage of these two classes of estimators and outline a semi-parametric approach to both methods by assuming that the distortion parameter $q = q(m)$ depends on the size of blocks m rather than the sample size n. We will proceed via simulation, addressing how the choice of q crosses over to the estimation of high quantiles, including the finite upper endpoint. The simulation study will be partially mirrored in the practical application to the annual maxima of Lowestoft sea levels.

C0218: **Semi-parametric estimation of the tail dependence coefficient through generalized means**

Presenter: **Ivette Gomes**, FCiencias.ID, Universidade de Lisboa and CEAUL, Portugal

Many examples in the most diverse fields of application show the need for statistical methods of analysis of extremes of multivariate data. And a crucial issue that appears when there is more than one variable is that of dependence. The study of multivariate extremes can be split essentially in two parts: the marginal distributions and the dependence structure. In first place, the margins are dealt with, and univariate extreme value theory techniques are used. In second place, we have to deal with dependence, also often dependent on univariate techniques, just as will be seen for the estimation of the tail dependence coefficient (TDC). Indeed, and thinking on a bivariate framework, (X,Y), after the standardization of the margins to a unit Fréchet distribution, the TDC appears as the reciprocal of the regularly varying exponent of a Pareto-type right-tail function of the differences between the margins. Different generalized means have been recently used in a successful estimation of the extreme value index, among other parameters of univariate extreme events, and will be now used for the TDC-estimation.

CO060 Room Cuza Hall SMALL AREA ESTIMATION: MODELS AND APPLICATIONS Chair: Maria Dolores Ugarte

C0171: **Combined analysis of misaligned data using Gaussian fields and the stochastic partial differential equation approach**

Presenter: **Paula Moraga**, Lancaster University, United Kingdom

Spatially misaligned data are becoming increasingly common. To optimize the use of these data, methods to combine data at different spatial scales and enable better predictions are needed. Current approaches present some limitations in terms of convergence and computational time. We present a geostatistical model for fusion of data obtained at point and areal resolutions using INLA and SPDE. This new approach is fast and flexible. The model presented assumes that underlying all observations there is a spatially continuous variable that can be modeled using a Gaussian random field process. In the SPDE approach, the continuously indexed Gaussian random field is represented as a discretely indexed Gaussian Markov random field (GMRF). To allow the combination of data, we propose a new projection matrix for mapping the GMRF from the observation locations to the triangulation nodes. The performance of the model is compared with the method RAMPS via simulation. The model is also applied to predict the concentration of fine particulate matter. The results show that the combination of point and areal data provides better predictions than if the method is applied to just one type of data, and this is consistent over both simulated and real data. The approach presented is a helpful advance in the area of spatial statistics that provides a useful tool that is applicable in a wide range of situations.

C0254: **Age, time, and gender-specific share component models to predict cancer incidence when mortality is known**

Presenter: **Jaione Etxeberria**, Public University of Navarre, Spain

Co-authors: Tomas Goicoa, Maria Dolores Ugarte

Updated incidence and mortality measures play an important role in a comprehensive overview of cancer burden. In Spain, cancer mortality figures are routinely recorded by Statistical Offices while cancer incidence is systematically recorded by regional cancer registries. Generally, incidence numbers become available three or four years later than mortality figures. In this context, to predict incidence rates in periods when the mortality is already known becomes necessary in order to provide the most updated cancer overview. According to International Cancer Agencies, realistic predictions of incidence rates should fulfil a list of requirements: 1-They should be stable over time, 2-They must be comparable in different populations or regions, 3-Age-specific incidence curves should be provided (including childhood cancer rates) and 4-Mortality-to-Incidence ratios should be taken into account. Considering all these, we propose to use age, time, and gender-specific shared component models for predicting incidence rates in lethal cancers where there exist a high correlation between incidence and mortality. Different models will be considered and their performance will be analyzed using brain cancer incidence and mortality data by gender and age-groups in 27 health units from Navarre and Basque Country (two Spanish regions) during the period 1998-2008. A fully Bayesian approach based on integrated nested Laplace approximations will be considered for model fitting and inference.

C0232: **The Mellin transform as a tool for prior elicitation in disease mapping**
Presenter: **Fedele Greco**, University of Bologna, Italy
Co-authors: Linda Altieri, Carlo Trivisano

Disease mapping encompasses a set of methodologies employed to describe the disease risk distribution over a study region. When the disease under study is rare, counts are heavily affected by random variability, and the estimates of the relative risk at the small-area level are unstable. The main aim of disease mapping studies is the identification of the underlying distribution of the risk. Several approaches have been proposed for modelling unstructured and spatially structured components. After a discussion about the existing proposals for prior specification in disease mapping, we present a novel approach that allows full control on prior specification for the well-known Besag-York and Mollie' model. Our proposal is built to the aim of matching the marginal variability of the structured and unstructured components, so that priors on such components express exactly the same belief in terms of variability allocation. The goal is reached by inversion of the Mellin transform of the distribution of a quadratic form. Both an application and a simulation study are presented for comparison with other proposals.

C0290: **On multivariate CAR models for multivariate disease mapping: The state of the art**
Presenter: **Ying C MacNab**, University of British Columbia, Canada

Multivariate disease mapping has been a very active area of research in recent decade. Notable progresses have been made in the formulation of flexible proper multivariate conditional autoregressive (MpCAR) models. One attractive feature of this class of MpCARs is that they can be formulated to model multivariate spatial dependencies, including (a)symmetric cross-dependencies and (a)symmetric cross-covariance functions. These MpCARs are typically used as disease risks priors within a Bayesian hierarchical inferential framework, to enable multivariate spatial smoothing and posterior risk prediction and inference. A brief survey of the recent proposals of MpCARs will be given, with a focus on enforcement for positive definiteness via constrained parameterization or reparameterization. Sufficient as well as sufficient and necessary constraints and related computational options are presented. They are the current state-of-the-art and are of central importance in posterior estimation and inference of Bayesian hierarchical models using MpCARs.

C0350: **Conditional autoregressive models for disconnected graph**
Presenter: **Anna Freni Sterrantino**, Imperial College London, United Kingdom
Co-authors: Haavard Rue, Massimo Ventrucci

Conditional autoregressive (CAR) distributions are widely used in disease mapping to created smoothed relative risk maps. The underlying maps are defined using an adjacency matrix based on the spatial neighborhood of areal units, de facto defining connected graphs, but in presence of islands or discontinue geographical regions disconnected graphs are created. Currently, if there are islands in the map (singletons) usually are connected to the closest areal unit or a constant prior is assigned for the singleton making it difficult for the singleton random effect to shrink to the global mean. Additionally, if the map has several connected the marginal deviation from its (component) mean, depends on the graph, and will, in general, be different for each connected component. To solve the issues we scale the precision matrix of the CAR, in detail we scale each connected component of size larger than one, independently and for connected components of size one, instead of setting the singleton random effect equal to zero, we replace it with a standard Gaussian with precision. This scaling gives a well-defined interpretation of the precision and the same typical (conditional) marginal variance within each connected component, no matter the size of the map.

CO040 Room Mezzanine Lounge **ADVANCES IN SURVIVAL AND RELIABILITY** **Chair: Juan Eloy Ruiz-Castro**

C0216: **A proportional hazards model under bivariate censoring and truncation**
Presenter: **Marialuisa Restaino**, University of Salerno, Italy
Co-authors: Hongsheng Dai

Bivariate survival data have received considerable attention recently. However, most existing research works have focused on bivariate survival analysis when one component is censored or truncated and the other is fully observed. Only recently bivariate survival function estimation when both components are censored and truncated has received considerable attention. In order to evaluate the incidence of covariates on the duration time, the proportional hazards model is used. The estimation of the regression coefficients in the Cox Proportional Hazards model is considered when the components are both censored and truncated. Moreover, we take into account that truncation could affect directly the hazard function. A simulation study is conducted to investigate the performance of the estimators of the unknown parameters.

C0302: **Reliability of particular semi-Markov systems: Modeling and nonparametric estimation**
Presenter: **Vlad Barbu**, Universite de Rouen, France

Three particular types of semi-Markov systems are considered which are associated reliability theory and nonparametric estimation. More precisely, we assume particular cases for the holding time distributions: they may depend only on the current state, or only on the next state to be visited, or neither on the current state, nor on the next state. These specific cases can be important from a practical point of view in some specific applications. We obtain explicit forms for the reliability indicators (reliability, availability, maintainability, failure rate, mean times, etc.) and propose consistent nonparametric estimators. Several estimation frameworks are taken into account: one or several trajectories, and complete or censored sample paths.

C0320: **Dependence model for complex deterioration systems**
Presenter: **Nuria Caballe Cervigon**, University of Extremadura, Spain
Co-authors: Inma Torres Castro

A condition-based maintenance (CBM) strategy is considered for a system subject to multiple degradation processes. Degradation processes start at random times following a Non-Homogeneous Poisson Process and the growth of these processes is modelled by using a gamma process whose parameters depend on the number of degradation processes existing in the system. It implies that the number of degradation processes in the system increases the speed of deterioration of the system. The system fails when the deterioration level of a degradation process exceeds a failure threshold. Additionally, the state of the system is inspected at different times and different maintenance tasks are performed in these inspection times. Usually, these inspection times are scheduled periodically. However, during the inspection, a lot of information about the status of the system is obtained. Such information could use to schedule the next inspection time. We consider that the inspection times are scheduled taking into account the number of degrading processes and their degradation levels. Numerical examples are provided to illustrate the optimisation of the objective function of the model defined by the expected cost rate.

C0303: **On dynamic modelling in reliability theory**

Presenter: **Alexandros Karagrigoriou**, University of The Aegean, Greece

Co-authors: Vlad Barbu, Andreas Makrides

The focus is on a general class of distributions for independent not necessarily identically distributed (inid) random variables, closed under extrema, that includes a number of discrete and continuous distributions like the Geometric, Exponential, Weibull or Pareto. The scale parameter involved in this class of distributions is assumed to be time varying with several possible modeling options proposed. Such a modelling setting is of particular interest in reliability and survival analysis for describing the time to event or failure. The maximum likelihood estimation of the parameters is addressed, and the asymptotic properties of the estimators are discussed. We provide real and simulated examples, and we explore the accuracy of the estimating procedure, as well as the performance of classical model selection criteria in choosing the correct model among a number of competing models for the time-varying parameters of interest.

C0255: **MMAPs to model a warm standby multi-state system with loss of units and a general variate number of repairpersons**

Presenter: **Juan Eloy Ruiz-Castro**, University of Granada, Spain

Co-authors: Mohammed Dawabsha

A complex warm standby multi-state system subject to different types of failures, repairable and non-repairable, is modeled through a Discrete Markovian Arrival Process with marked arrivals (D-MMAP). The system is composed of a general number of units, K, and RK repairpersons. Each unit can undergo a failure at any time. The online unit failure can be internal (repairable or non-repairable) due to wear or external due to one external shock. When one external shock occurs, it can provoke a modification in the internal behavior of the online unit or a fatal failure. Each warm standby system can undergo a repairable failure. A non-repairable failure implies that the unit is removed and in this case the system continues working with a less unit while it is possible. If it occurs, the number of repairpersons is also modified. Some interesting measures, such as the mean operational time and the mean number of events up to a given time have been worked out. Costs and rewards are included in the model and the expected cost up to a certain time is calculated. The number of repairpersons has been optimized according to the number of units in the system. The model is built in an algorithmic form which eases the computational implementation.

CO050 Room Vega Hall JCS SESSION: VISUALIZATION OF A HIGH DIMENSIONAL DATA MATRIX Chair: Tadashi Imaizumi

C0317: **Visualization of disaster information over time by using disaster information extraction from Twitter**

Presenter: **Yoshiro Yamamoto**, Tokai Univeristy, Japan

Co-authors: Takamitsu Funayama, Osamu Uchida

Many natural disasters such as earthquakes occurred in Japan. Many cases have been reported that use SNS such as Twitter to reduce disaster in the event of a disaster. We aim to extract useful information for grasping the situation at the time of disaster from Twitter and to visualize the information to make it easier to understand. Since Twitter information is Tweeted freely by unspecified majority in the whole world, it is difficult even by limiting specific disasters to Tweet. It is necessary to extract Tweet extracted by keywords such as "earthquake", Tweet by the person actually affected has occurred. In visualization that can grasp the situation of the earthquake by text mining, visualization was devised aiming at grasping the change of the situation in consecutive several hours unit. We analyzed information posted on Twitter during the earthquake in Kumamoto in April 2016. We will report a case where we visualize the change of Tweet for two days from the occurrence of the earthquake.

C0288: **Visualization and spatial statistics for spatial data that contribute to comprehensive suicide countermeasures in Japan**

Presenter: **Takafumi Kubota**, Tama University, Japan

Co-authors: Marina Ishikawa, Yoshikazu Yamamoto

Spatial data of suicides in Japan are visualized, which contribute to the development of comprehensive suicide countermeasures. To observe the suicide risks based on area, age, and gender, small area suicide data (suicide data) are visualized. These data are derived on the premise that the place is determined based on the location at which a person commits suicide, and such locations are identified by observing the "mobile spatial statistics". To estimate the actual relation between industrial structures and suicide risks, the suicide data are visualized based on the principle that the place of suicide is based on the residence of the person who has committed suicide; the required information can be obtained using the RESAS (Regional Economy and Society Analyzing System) data. Suicide data are also used to conduct spatial autocorrelation and conditional autoregressive modeling. To detect spatial autocorrelation, a local Moran plot of the suicide data for visualization is created. Further, a regression model is used based on several variables, such as health, income, tax, education, and so on, to identify the risks that are associated with various regions.

C0271: **Cluster detection for multi-dimensional spatial data based on hierarchical structure**

Presenter: **Fumio Ishioka**, Okayama University, Japan

Co-authors: Koji Kurihara

In recent years, the spatial scan statistics that detect a spatial cluster for spatial data is widely used in spatial epidemiology and other fields. However, currently, most of them are limited to applications on two-dimensional data such as geospatial data. Echelon analysis is an approach which enable us to visualize the spatial data systematically and objectively by a topological hierarchical structure according to the adjacency relationship of each region. Even if each region has multiple variables, if they are ordinal variables, and we can define relative positions between variables based on their order, it is possible to draw them as a two-dimensional dendrogram. Therefore, the echelon scan method that combines a spatial scan statistic and echelon analysis enables us to detect a spatial cluster for spatial data with multidimensional nature. We introduce how to perform the cluster detection for spatial data having multidimensional elements such as time series or multivariate by using the echelon scan method with showing a concrete example.

C0282: **The classification and visualization of trending topics in online word-of-mouth data**

Presenter: **Atsuho Nakayama**, Tokyo Metropolitan University, Japan

Trending topics in online word-of-mouth data are classified by focusing on topics related to new products. The analysis of large amounts of online word-of-mouth data collected from Social Networking Service has received much attention to help identify market trends. Twitter has been widely using in Japan. Consumers post a lot of comments regarding a wide range of topics in Twitter. The tweets include opinions about products and services. We collected Twitter entries about new products based on their specific expressions of sentiment. We tokenized each tweet message that was written in sentences or sets of words to detect topics more easily. Morphological analyses has been that such as tokenization, stemming, and part-of-speech tagging to separate the words. Next, we selected keywords representative of our chosen topics. We performed a statistical analysis based on the complementary similarity measure that has been widely applied in the area of character recognition. Then, we detected trending topics related to a new product by classifying words into clusters based on the co-occurrence of words in Twitter entries. Topic-based sentiment analysis has been used to extract and visualize the topic of customer interests and opinions.

C0296: **Clutering in reduced space of high-dimensional binary data**

Presenter: **Tadashi Imaizumi**, Tama University, Japan

In Behavioral Science, or Marketing Science etc, we need to analyze a binary data matrix whose cells represent whether a respondent responded to an item or not. When this binary data matrix has few factors, less than 30, Multiple Correspondence Analysis (MCA) or Joint Correspondence Analysis (JCA) have been applied, and they are useful in analyzing Bart tables. When the number of factors is larger, for example, 100 or 200, it will be hard to understand the derived results. So, a new method will be proposed for analyzing this type of binary data. In this method, an unknown number of G clusters are assumed. The response on each binary variable is also assumed to be connected with the distance between a cluster center and the center point on the dimension representing that variable. Each of the G cluster centers will be embedded as a point in a lower dimensional space, typically, a 2-dimensional space. They are derived by the orthonormal rotation of the cluster center in a higher dimensional space. This orthonormal matrix will be derived by maximized the between variance among the cluster centers in the lower dimensional space. An application of the proposed method to a real data set will be shown.

CC078 **Room Clio Hall** **ALGORITHMS AND COMPUTATIONAL METHODS** **Chair: Woncheol Jang**

C0334: **Efficient Monte Carlo evaluation of resampling-based hypothesis tests**
Presenter: **Wing Kam Fung**, University of Hong Kong, Hong Kong

Monte Carlo evaluation of resampling-based tests is often conducted in statistical analysis. However, this procedure is generally computationally intensive. The pooling resampling-based method has been developed to reduce the computational burden but the validity of the method has not been studied before. The asymptotic properties of the pooling resampling-based method are first investigated. A novel Monte Carlo evaluation procedure namely the n-times pooling resampling-based method is then proposed. Theorems as well as simulations show that the proposed method can give smaller or comparable root mean squared errors and bias with much less computing time, thus can be strongly recommended especially for evaluating highly computationally intensive hypothesis testing procedures.

C0277: **Automatic anomaly detection in jet engines**
Presenter: **Harjit Hullait**, Lancaster University, United Kingdom
Co-authors: David Leslie, Nicos Pavlidis

Jet Engines contain hundreds of sensors measuring various engine processes in continuous time. Building sophisticated statistical methods to process and make informative decisions from the sensor data offers huge opportunities. We will start by looking at sensor data from Pass-Off tests. Each test comprised an engine performing various pre-defined manoeuvres. There are two types of manoeuvres: piecewise-linear manoeuvres, containing sudden changes in behaviour and functional manoeuvres, which are comprised of smooth accelerations and decelerations. There are a number of challenges in using this data; firstly, the manoeuvres performed have not been labelled and secondly, manoeuvres can be partially performed. The first aim is to build an efficient classification method to identify the different manoeuvres. We have used the PELT changepoint algorithm to extract manoeuvre segments. We then use Needleman-Wunsch and Functional Principal Component Analysis to compare the manoeuvre segments to a number of model templates, which gives us a similarity score with respect to each template. We use the Needleman-Wunsch algorithm, as it is capable of identifying missing segments, so is particularly robust in scoring partially performed manoeuvres. These scores characterise the manoeuvres effectively, enabling the manoeuvres to be classified with a high level of accuracy.

C0337: **Acceleration of computation for fuzzy c-means clustering**
Presenter: **Yuichi Mori**, Okayama University of Science, Japan
Co-authors: Takatsugu Yoshioka, Masahiro Kuroda

Fuzzy c-means clustering (FCM), which is a nonhierarchical and soft clustering method, sometimes requires high computational cost due to the iterative convergence in the computation. To reduce the cost, a general procedure to accelerate the iterative computation such as alternating least squares has been proposed. This procedure generates a new accelerated convergent sequence using the vector epsilon algorithm based on the original convergent sequence in estimating two or more parameters alternatingly. Since the procedure can be applied to any computation which generates a linearly convergent sequence, it is applied to FCM, in which the membership matrix and the cluster centroid matrix are estimated alternatingly until convergence, to obtain the computational results faster than the original computation. Some numerical experiments demonstrate that the vector epsilon accelerated FCM accelerates the computation twice faster or more than the original one.

C0355: **fiberLD: An R package for fiber length determination**
Presenter: **Natalya Arnqvist**, Umea University, Sweden
Co-authors: Konrad Abramowicz, Sara Sjostedt de Luna

Methods for estimating tree fiber (tracheid) length distributions in the standing tree based on increment core samples have been implemented in an R package fiberLD. Two types of data can be used with the package, increment core data measured by means of an optical fiber analyzer (OFA), e.g. such as the Kajaani Fiber Lab, or measured by microscopy. Increment core data analyzed by OFAs consist of the cell lengths of both cut and uncut fibres (tracheids) and fines (such as ray parenchyma cells) without being able to identify which cells are cut or if they are fines or fibers. The microscopy measured data consist of the observed lengths of the uncut fibers in the increment core. A censored version of a mixture of the fine and fiber length distributions is proposed to fit the OFA data, under distributional assumptions. The package offers two choices for the assumptions of the underlying density functions of the true fiber (fine) lengths of those fibers (fines) that at least partially appear in the increment core, being the generalized gamma and the log normal densities.

C0385: **Computing p-values of the Kolmogorov-Smirnov test for (dis)continuous null distribution: R package KSgeneral**
Presenter: **Senren Tan**, Cass Business School, City, University of London, United Kingdom
Co-authors: Dimitrina Dimitrova, Vladimir Kaishev

The distribution of the Kolmogorov-Smirnov (K-S) test statistic has been widely studied under the assumption that the underlying theoretical cumulative distribution function (cdf), F(x), is continuous. However, there are many real-life applications in which fitting discrete or mixed distributions is required. Nevertheless, due to inherent difficulties, the distribution of the K-S statistic when F(x) has jump discontinuities has been studied to a much lesser extent and no exact and efficient computational methods have been proposed in the literature. A fast and accurate method to compute the (complementary) cdf of the K-S statistic when F(x) is discontinuous is provided, and thus exact p-values of the K-S test are obtained. The approach is to express the complementary cdf through the rectangle probability for uniform order statistics, and to compute it using Fast Fourier Transform (FFT). Secondly, a C++ and an R implementation of the proposed method are provided, which fills in the existing gap in statistical software. The numerical performance of the proposed FFT-based method, implemented both in C++ and in the R package KSgeneral, is illustrated when F(x) is mixed, purely discrete, and continuous.

CC079 **Room Orion Hall** **MULTIVARIATE METHODS** **Chair: Maria-Pia Victoria-Feser**

C0281: **Crucial differences between principal component and factor analyses solutions elucidated by some inequalities**
Presenter: **Kohei Adachi**, Osaka University, Japan
Co-authors: Nickolay Trendafilov

Some inequalities are presented to unmask the differences between the principal component analysis (PCA) and factor analysis (FA) solutions for the same data set. For this reason, we take advantage of the matrix decomposition (MD) formulation of FA established recently. In summary, the resulting inequalities show that [1] FA provides a better fit to the data than PCA, [2] PCA extracts a larger amount of common information than

FA, and [3] For each variable, its unique variance in FA is larger than its residual variance in PCA minus the one in FA. The resulting inequalities can be useful to suggest whether PCA or FA should be used for a particular data set. The answers can also be valid for the classic random FA not relying on the MD-FA formulation, as both types of FA are found to provide almost equal solutions. Additionally, the inequalities give theoretical explanation of some empirically observed tendencies in PCA and FA solutions, e.g., that the absolute values of PCA loadings tend to be larger than those for FA loadings, and that the unique variances in FA tend to be larger than the residual variances of PCA.

C0338: **Maximum likelihood EM factor analysis of high-dimensional data with the sparsest constraint on loadings**

Presenter: **Jingyu Cai**, Osaka University, Japan

Co-authors: Kohei Adachi

The constrained factor analysis (FA) procedures have been proposed, in which each variable is constrained to load only one factor. Thus, the resulting loading matrix is the sparsest, in that it has a single nonzero element in each row and zeros elsewhere. Such procedures can be called sparsest FA. We propose a new sparsest FA procedure feasible for the high-dimensional data with the number of variables greater than that of observations. Such data have not been considered in the existing approaches. In the proposed procedure, the FA log-likelihood is maximized over loadings, factor correlations, and unique variances, with the loadings constrained to be the sparsest. This maximization is attained using a modified version of the EM algorithm for confirmatory FA. The original EM algorithm for FA is feasible to high-dimensional data and our modified version also has this property. The loadings being the sparsest facilitates their interpretation, in particular, for high-dimensional cases, as a great number of variables are classified exclusively into a few clusters on the basis of what factor is loaded by each variable. Such a benefit is illustrated with numerical examples.

C0386: **Probability of success for regulatory decision-making in a pediatric indication**

Presenter: **Aiesha Zia**, Novartis Pharma, Switzerland

Co-authors: Simon Wandel, Nathalie Fretault, Shantha Rao

Two formulations of a marketed drug are investigated in a pediatric study. The primary objective is to evaluate treatment efficacy in the respective population of interest. Recruitment is particularly challenging, imposing a risk to meet regulatory requirements. An early analysis based on partially complete data was prepared for strategic discussion. This strategy involved probability of success (PoS) calculation to enrich the tools available for decision-making. The primary endpoint was a continuous longitudinal endpoint assessed at a specific time point. Several multivariate models, including a Bayesian approach with covariance matrix under different assumptions, to predict the yet unobserved data were used for PoS calculations. All models were fitted using SAS proc MI and proc MCMC. The PoS from general models properly reflects uncertainty due to missing values. Difficulties in fitting models were encountered due to sparse data, these challenges could be addressed by the models with more structural constraints. While PoS is a useful metric for decision-making, special considerations need to be given when data are sparse. The framework is particularly relevant to estimate PoS in the context of longitudinal partially complete data.

C0404: **Seeking kinks in a protein alpha-helix**

Presenter: **Mai Alfahad**, Leeds university, United Kingdom

Proteins are biomolecule compounds that consist of a chain of amino acid residues, which are of particular importance as they can be found in every living organism. The shape of protein plays an importance role in its function. The chain of amino acid form 3-dimensional shapes, of these, we are interested only in the common shape, Alpha-helix. The Alpha-helix is a right-handed helix. We study the shape of a helix and, more generally, a helix with kinks (change point), since kinks are functionally important in membrane proteins. Kinks are points where the helix axis changes direction. We developed a new algorithm to fit an unkinked helix, OptLS, for the maximum likelihood estimation (MLE). In addition, we establish a method to determine if a given protein Alpha-helix is kinked or not; and if it is kinked, then we find the kink position. If the helix in known to be kinked, then the kink position can be found, so that we can fit each unkinked half of the helix individually. In addition, we study 6 test statistics to investigate the nature of a kink.

C0411: **A pseudo-likelihood approach for multivariate meta-analysis of test accuracy studies with multiple thresholds**

Presenter: **Duc Khanh To**, University of Padova, Italy

Co-authors: Annamaria Guolo

In meta-analysis of test accuracy studies, the multivariate approach is an effective technique to synthesize results when each study reports sensitivities and specificities at different thresholds. The normal multivariate mixed-effects model proposed in literature as an extension of the well-known bivariate model for the one threshold case, despite interesting features, suffers from some drawbacks. They include the requirement of an estimate of within-study correlations between sensitivities (and specificities) at different thresholds and convergence issues. In order to overcome such drawbacks, we propose a pseudo-likelihood approach under a working independence assumption between sensitivities (and specificities) at different thresholds in the same study. The approach does not require the within-study correlations to be known or estimated and convergence issues very rarely occur. In addition, its implementation is straightforward. The problem of different set of thresholds per study is taken into account by assuming that the thresholds in each study are missing completely at random and adopting an available case. Several simulation studies show that the proposed method performs satisfactorily and improves on the corresponding results from the normal multivariate mixed-effects model. In order to illustrate the applicability of the pseudo-likelihood approach, some published meta-analyses of diagnostic test accuracy have been analyzed.

Thursday 30.08.2018 14:15 - 15:45 Parallel Session J – COMPSTAT2018

CI010 Room Cuza Hall COMPUTATIONAL DEVELOPMENTS FOR DISCRETE DATA Chair: Dimitris Karlis

C0209: **On modelling multivariate time series of counts**
Presenter: **Dimitris Karlis**, RC Athens University of Economics and Business, Greece

Multivariate count models appear in many disciplines, like marketing, epidemiology, seismology just to name few. In many cases the data are time series, i.e. multiple counts observed in a sequel of time points. Hence, for correct modelling we need to account for both serial and cross correlations. While the literature on continuous time series abandons, discrete valued time series have been given less interest. There is an increasing literature and number of models in recent years. For multivariate count time series thing are even more sparse. Observation driven models for the univariate case are now quite popular. We have previously introduced and extended such models in the multivariate case offering different approaches for model building and estimation. The aim is to introduce such models and provide some new results related to this family of models, including model selection and computational approaches to improve the performance of multivariate autoregressive models. Real data applications will be also described from different disciplines.

C0444: **Computational tools for count data regression**
Presenter: **Christian Kleiber**, Universitaet Basel, Switzerland

An overview of a variety of computational tools for count data regressions that are available via the R package **countreg** is provided. The package provides a number of fitting functions and new tools for model diagnostics: it incorporates enhanced versions of fitting functions for hurdle and zero-inflation models that have been available via the **pscl** package for some 10 years, now also permitting binomial responses. In addition, it provides zero-truncation models for data without zeros, along with **mboost** family generators that enable boosting of zero-truncated and untruncated count data regressions, thereby supplementing and extending family generators available with the **mboost** package. For visualizing model fits, **countreg** offers rootograms and probability integral transform (PIT) histograms. A (generic) function for computing (randomized) quantile residuals is also available. Development versions of **countreg** can be obtained from R-Forge. Some well-known data sets from the count data literature will be revisited.

CO024 Room Cocktail Hall TEXT MINING IN ECONOMICS AND FINANCE Chair: Peter Winker

C0308: **Identifying firm-level news shocks from financial news media**
Presenter: **Julian Ashwin**, Nuffield College, University of Oxford, United Kingdom

A text mining approach is investigated which can be used to identify firm level news shocks from news media, and whether these are of macroeconomic relevance. We use a variety of approaches, including Named-entity recognition, to match articles from the Financial Times newspaper to firms listed on the London Stock Exchange. We find that being mentioned in that day's edition has a robust, statistically significant and substantial effect on both absolute return and trading volume of an individual firm's stock price. Both supervised and unsupervised topic modelling approaches are used to extract richer information from news media which can predict firm level stock returns and separate "good" and "bad" news. We argue that different topics plausibly identify not only shocks to sentiment, but also new information about a firm's future profitability. This also allows the comparison of the effects of different types of news on an individual firm's stock price. These identified news shocks are then used to investigate how new information propagates across the stock market, and whether this process is related to the structure of the production network. This evidence is used to assess the plausibility of the claim that firm or sector specific news shocks could have effects similar to those of a negative technology shock by resulting in a misallocation of production factors.

C0269: **Analyzing economic texts using network based topic models**
Presenter: **Ryohei Hisano**, University of Tokyo, Japan

Topic models are one of the most commonly used statistical modelling approaches when analyzing economic and financial texts, due to its high interpretability and efficient nature. However, when analyzing economic texts, basic information besides the good old bag-of-words matrix, that topic models aim to model, becomes important. These additional information includes information such as who mentioned what at what timing, what were the economic indicators when a text was written, or even simply the meaning of a word to just name a few. We show a simple network based approach to incorporate these additional information into a topic model and observe what additional insights could be gained from our approach. We would mainly focus on analyzing the Economy watcher survey, published by the Cabinet office of Japan, but other data sets might be mentioned if time permits.

C0219: **Measuring the diffusion of innovations with paragraph vector topic models**
Presenter: **David Lenz**, Justus-Liebig University Giessen, Germany
Co-authors: Peter Winker

Topic modeling became an intensively researched area lately, mainly due to the ever-increasing availability of huge digital text information and the improvements in methods to analyze these datasets. In natural language processing, topic modeling describes a set of methods to extract the latent topics from a collection of documents. Several new methods have recently been proposed to improve the topic generation process. However, examination of the generated topics is still mostly based on unsatisfactory practices, for example by looking only at the list of most frequent words for a topic. Our contribution is threefold: 1) We present a topic modeling approach based on neural embeddings and Gaussian mixture modeling, which is shown to generate coherent and meaningful topics. 2) We propose a novel "topic report" based on dimensionality reduction techniques and model generated document vector features which helps to easily identify topics and significantly reduces the required mental overhead. 3) Lastly, we demonstrate on a technology related newsticker corpus how our approach could be used by economists to tackle economic problems, for example to measure the diffusion of innovations.

C0333: **Using large and heterogeneous sources of sentiment and attention data for predicting stock market volatility**
Presenter: **Fabio Sigrist**, Lucerne University of Applied Sciences, Switzerland
Co-authors: Daniele Ballinari, Francesco Audrino

The impact of sentiment and attention variables on volatility is analyzed by using a novel and extensive dataset that combines social media, news article, information consumption, and search engine data. Applying a state-of-the-art sentiment classification technique, we investigate whether sentiment and attention variables contain additional predictive power for realized volatility when controlling for a wide range of economic and financial predictors. Using a penalized regression framework, we identify investors' attention, as measured by Google searches about financial keywords (e.g. "financial market" and "stock market"), and the daily volumes of company-specific messages posted on StockTwits to be the most relevant variables. In addition, it is shown that attention and sentiment variables are able to significantly improve volatility forecasts, although the improvements are of relatively small magnitude from an economic point of view.

CO090 Room Mezzanine Lounge ADVANCES IN STRUCTURAL EQUATION MODELING AND PLS PATH MODELING Chair: Laura Trinchera

C0312: **SO-PLS-PM: From multiblock data analysis to path modeling**
Presenter: **Rosaria Romano**, University of Calabria, Italy
Co-authors: Oliver Tomic, Kristian H Liland, Age Smilde, Tormod Naes

A new approach to path modeling named SO-PLS path modeling (SO-PLS-PM) is presented and compared with the more well-known PLS path modeling (PLS-PM). The new method is flexible, graphically-oriented, and allows for handling multidimensional blocks and diagnosing missing paths. Instead of fitting everything at the same time using the full path model scheme, the approach splits the estimation up into separate multi-block regression models for each dependent/endogenous block. In other words, thepath modeling is turned into a series of regression analyses. Since the whole procedure is based on PLS regression and orthogonalization, the method can be used for any dimensionality of the blocks, for collinear variables as well as design data and it is invariant to the relative weighting of the blocks. In order to allow for a thorough comparison between the two methods, new definitions of total, direct and indirect effects in terms of explained variances are proposed, along with new methods for graphical representation. The two PLS methods are tested on two well-known data sets in the PLS-PM literature from customer satisfaction analysis and descriptive sensory analysis. The findings from the empirical applications serve as a basis for recommendations and guidelines regarding the use of the SO-PLS-PM versus PLS-PM.

C0242: **Testing design-oriented auxiliary theories using PLS path modeling**
Presenter: **Florian Schuberth**, University of Twente, Netherlands
Co-authors: Joerg Henseler
Auxiliary theories are indispensable to operationalize abstract concepts in structural equation modeling (SEM). While for behavioral concepts, such as traits or attitudes, auxiliary theories already exist, the current literature lacks of auxiliary theories for design concepts. Obviously, this is particularly disadvantageous for disciplines that investigate design concepts or an interplay of design and behavioral concepts. To fill this gap, an auxiliary theory for design concepts is provided, i.e., how design concepts can be translated into constructs and observed variables. In doing so, the composite is introduced as a way to model design constructs - so-called artifacts - in SEM. To estimate structural equation models containing composites, the study at hand employs partial least squares path modeling (PLS-PM), a variance-based estimator to SEM. To assess the overall model fit in order to obtain empirical evidence for the built composites, a bootstrap-based testing procedure is applied. It compares the distance between the empirical and the model-implied variance-covariance matrix of the indicators to a critical value from its corresponding reference distribution in order to draw conclusions about the usefulness of the artifacts. This allows for answering new research questions of the sort "Is the built construct useful?"

C0313: **A new confindence interval for Cronbachs coefficient alpha**
Presenter: **Laura Trinchera**, NEOMA Business School, France
Co-authors: Nicolas Marie, George Marcoulides
Reliability is commonly examined in order to assess the measurement quality of scales. To date, Cronbachs coefficient alpha is the most commonly used index for assessing the reliability of a scale. We present an asymptotic distribution of the natural estimator of Cronbachs alpha coefficient and propose a new CI that does not require assumptions of equal variances and covariances, and neither does it require the data to be approximated by a multivariate normal distribution. We also present a test on the sample estimate of coefficient alpha for testing the null hypothesis that the coefficient is higher than 0.7. The proposed approach is compared to four popular methods commonly used to compute confidence intervals (CI) for alpha using a Monte Carlo simulation study under a variety of sample size and number of items in a scale conditions. We compare results for each method in terms of the level of coverage and confidence interval length (CIL). The results of this simulation study indicated that the newly proposed interval estimate was the most accurate of the examined approaches, especially for small sample sizes.

CO068 **Room Vega Hall** **SOFT CLUSTERING** **Chair: Maria Brigida Ferraro**

C0208: **Fuzzy clustering of multivariate skew data**
Presenter: **Francesca Greselin**, University of Milano Bicocca, Italy
Co-authors: Luis Angel Garcia-Escudero, Agustin Mayo-Iscar
With the increasing availability of multivariate datasets, asymmetric structures in the data ask for more realistic assumptions, with respect to the incredibly useful paradigm given by the Gaussian distribution. Moreover, in performing ML estimation we know that a few outliers in the data can affect the estimation, hence providing unreliable inference. Challenged by such issues, more flexible and solid tools for modeling heterogeneous skew data are needed. Our fuzzy clustering method is based on mixtures of Skew Gaussian components, endowed by the joint usage of impartial trimming and constrained estimation of scatter matrices, in a modified maximum likelihood approach. The algorithm generates a set of membership values, that are used to fuzzy partition the data set and to contribute to the robust estimates of the mixture parameters. The new methodology has been shown to be resistant to different types of contamination, by applying it on artificial data. A brief discussion on the tuning parameters has been developed, also with the help of some heuristic tools for their choice. Finally, synthetic and real dataset are analyzed, to show how intermediate membership values are estimated for observations lying at cluster overlap, while cluster cores are composed by observations that are assigned to a cluster in a crisp way.

C0299: **Soft clustering-based models**
Presenter: **Mika Sato-Ilic**, University of Tsukuba, Japan
Researchers in the area of clustering data deal with large amounts of complex data. Soft clustering has the merit of explaining this data using a smaller number of clusters which enables us to obtain stable solutions in the sense of robustness and reproducibility. Model-based clustering is a framework of clustering methods that assume a model to the data so an adjusted partition can be estimated. Although this approach can obtain a clear solution as the result of the partition based on mathematical theory, we cannot avoid the risk that the assumed model might not adjust to the latent classification structure of the data. Therefore, we propose a framework called clustering-based models which exploits obtained clustering result as a scale of the latent structure of the data and applies it to the observed data. Since the cluster-based scale is obtained from the original data, the scale can measure the original data and then the re-measured data can be applied to the model to obtain a more accurate result. Soft clustering is utilized to capture the cluster-based scale to deal with the large amount of complex data and several soft clustering-based models with applications will be introduced.

C0427: **A social network analysis of articles on social network analysis**
Presenter: **Clement Lee**, Newcastle University, United Kingdom
Co-authors: Darren Wilkinson
A collection of articles on the statistical modelling and inference of social networks is analysed in a network fashion. The references of these articles are used to construct a citation network data set, which is (almost) a directed acyclic graph (DAG) because only existing articles can be cited. A mixed membership stochastic block model (MMSBM) is then applied to this data set to soft cluster the articles. For inference, both the regular and collapsed Gibbs samplers are presented with their performances compared. The results give us insights on the influence and categorisation of these articles.

C0300: **Relational fuzzy clustering to identify non-linear structures**
Presenter: **Maria Brigida Ferraro**, Sapienza University of Rome, Italy
Co-authors: Paolo Giordani
Classical (hard or fuzzy) algorithms usually detect clusters by computing the Euclidean distance among pairs of objects. They are based on the linearity assumption and, therefore, do not identify properly clusters characterized by non-linear structures. In order to overcome this limitation, the

so-called geodesic distance can be considered, where the linearity assumption holds locally. In fact, the geodesic distance between two neighboring objects is equal to the Euclidean one whilst, in case of two faraway objects, it is equal to the shortest path in the graph connecting them. The aim is to propose some relational fuzzy clustering methods for non-linear data.

CG063 **Room Orion Hall** ADVANCES ON THE ANALYSIS OF LARGE DATA SETS **Chair: Asaf Weinstein**

C0173: **Joint estimation of multiple penalized graphs using zoom-in/out penalties to map functional brain connectivity**

Presenter: **Eugen Pircalabelu**, KU Leuven, Belgium

Co-authors: Gerda Claeskens, Lourens Waldorp

A new method is proposed to simultaneously estimate graphical models from data obtained at K different coarseness scales. Starting from a predefined scale $k^* \leq K$, the method offers the possibility to zoom in or out over scales on particular edges. The estimated graphs over the different scales have similar structures, although their sparsity level depends on the scale at which estimation takes place. The graphs are jointly estimated at all coarseness scales and the method makes it possible to evaluate the evolution of the graphs from the coarsest to the finest scale or vice-versa. The method is motivated by fMRI datasets that do not all contain measurements on the same set of brain regions. For certain datasets, some of the regions have been split in smaller subregions and the purpose is to estimate sparse graphical models. We accomplish this by pooling information from all subjects in order to estimate a common undirected and directed graph at each coarseness scale, accounting for time dependencies and multiple coarseness scales and by jointly estimating the graphs at all coarseness scales. Empirical and theoretical evaluations illustrate the usefulness of the method and show the method's performance in practice.

C0364: **Normalizations in derived networks**

Presenter: **Vladimir Batagelj**, IMFM, Slovenia

Linked networks are collections of networks over at least two sets and consist of some one-mode networks over single sets and some two-mode networks linking them. A very important role in analysis of linked networks plays the network multiplication that enables us to produce so called "derived networks". For example: a two-mode network **PA** describes the authorship relation linking papers P to their authors A; and **Ci** is a one-mode citation network. Both networks are linked because they share the set of papers P. Let **AP** denote a network obtained from **PA** by reversing directions of all its links. Then the network $\mathbf{AP} * \mathbf{Ci} * \mathbf{PA}$ describes citations between authors from A. The weight on a link tells us how many times the first author cited in his/her papers the second author. Because large networks are usually sparse (the number of links is of the same order as the number of nodes) it is, in most cases, possible to compute their product fast. Using so called "fractional approach" – normalizing some matrices in the product we get different weights (with different meaning) in the derived network. We present a theoretical background of normalization in computing derived networks and illustrate the results with analyses of selected bibliographic networks.

C0414: **Posterior probability SVMs for big data problems**

Presenter: **Pedro Duarte Silva**, Universidade Catolica Portuguesa / Porto, Portugal

Assume that objects belonging to two well-defined groups can be described by random pairs (x,y), where x is an object descriptor and y is a group label. Then, given a training sample of l independent examples drawn from a common distribution, and a rich enough Reproducing Kernel Hilbert Space, the minimal misclassification probability Bayes rule can be consistently estimated by standard two-group classification Support Vector Machines (SVMs). However, if misclassification costs differ across groups or the training sample proportions do not converge to true a priori probabilities, standard SVMs do not approximate optimal Bayes rules. Nevertheless, non-standard SVMs that minimize a weighted misclassification loss plus a regularization penalty are consistent estimators of the minimal expected cost rule. Furthermore, posterior probabilities may be consistently estimated from the solutions of a succession of non-standard SVMs with varying weights in the loss function. A l1-norm regularization posterior probability SVM specially adapted for Big Data problems will be described. It is known that l1-norm based SVMs are able to handle efficiently much larger data sets than l2-norm based SVMs. The computational and statistical properties of this proposal will be illustrated by simulation experiments.

C0437: **A new noise-resisting feature-based clustering quality evaluation approach scaling from low to high dimensional data**

Presenter: **Jean-Charles Lamirel**, LORIA, France

The main concern is the optimal model selection in clustering. New quality indexes based on feature maximization are presented for that purpose. Feature maximization is an efficient alternative approach for feature selection in high dimensional spaces to usual measures like Chi-square, vector-based measures using Euclidean distance, correlation or information gain. The behavior of the new feature maximization based indexes is compared with a wide range of usual quality indexes, and with large set of alternative indexes as well, on different kinds of real life datasets constituted from low to high dimensional data for which ground truth is available. This comparison highlights the better accuracy and stability of the new indexes on these datasets, their efficiency from low to high dimensional range and their high tolerance to noise. Additional experiments are done on real life high dimensional textual data issued from a bibliographic database for which ground truth is unavailable. Experiments highlight that the accuracy and stability of these new indexes allow to efficiently manage time-based diachronic analysis. Conversely, usual indexes do not fit the requirements for this task. The proposed indexes are tested with hard clustering but their straightforward adaptation for soft clustering is finally presented.

Thursday 30.08.2018 16:15 - 17:45 Parallel Session K – COMPSTAT2018

CI115 Room Cuza Hall MODEL BASED CLUSTERING AND CLASSIFICATION **Chair: Francesca Greselin**

C0360: **Gaussian-based visualization of Gaussian and non-Gaussian model-based clustering**
Presenter: **Christophe Biernacki**, Inria, France
Co-authors: Vincent Vandewalle, Matthieu Marbac-Lourdelle

A generic method is introduced to visualize in a Gaussian-like way, and onto *Rd*, results of Gaussian or non-Gaussian model-based clustering. The key point is to explicitly force a spherical Gaussian mixture visualization to inherit from the within cluster overlap which is present in the initial clustering mixture. The result is a particularly user-friendly draw of the clusters, allowing any practitioner to have a thorough overview of the potentially complex clustering result. An entropic measure allows us to inform of the quality of the drawn overlap, in comparison to the true one in the initial space. The proposed method is illustrated on four real data sets of different types (categorical, mixed, functional and network) and is implemented on the R package ClusVis.

C0367: **The role of trimming and variable selection in robust model-based classification for food authenticity studies**
Presenter: **Andrea Cappozzo**, University of Milano Bicocca, Italy
Co-authors: Francesca Greselin, Thomas Brendan Murphy

Food authenticity studies deal with the detection of products that are not what they claim to be, thereby preventing economic fraud or possible damage to health. For identifying illegal sub-samples we introduce robustness in a semi-supervised model-based classification rule. That is, labelled and unlabelled data are jointly modeled by a Gaussian mixture model with parsimonious covariance structure. To avoid singularity issues, we adopt a restriction on the eigenvalues'ratio of the group scatter matrices. Adulterated observations are detected by monitoring their contributions to the overall observed likelihood, and following the impartial trimming established technique: the illegal sub-sample is the least plausible under the estimated model. A wrapper approach for variable selection is then considered, providing relevant information about discriminant variables and for feature reduction in a high-dimensional context. Experiments on real data, artificially adulterated, are provided to underline the benefits of the proposed method.

C0396: **Kernel-estimated nonparametric overlap-based syncytial clustering**
Presenter: **Ranjan Maitra**, Iowa State University, United States
Co-authors: Israel Almodovar Rivera

Standard clustering algorithms usually find regular-structured clusters such as ellipsoidally- or spherically-dispersed groups, but are more challenged with groups lacking formal structure or definition. Syncytial clustering is the name that we introduce for methods that merge groups obtained from standard clustering algorithms in order to reveal complex group structure in the data. We develop a distribution-free fully-automated syncytial clustering algorithm that can be used with k-means and other algorithms. Our approach computes the cumulative distribution function of the normed residuals from an appropriately fit k-groups model and calculates the nonparametric overlap between each pair of groups. Groups with high pairwise overlaps are merged as long as the generalized overlap decreases. Our methodology is always a top performer in identifying groups with regular and irregular structures in several datasets. The approach is also used to identify the distinct kinds of gamma ray bursts in the Burst and Transient Source Experiment 4Br catalog and also the distinct kinds of activation in a functional Magnetic Resonance Imaging study.

CO042 Room Mezzanine Lounge BAYESIAN STATISTICS **Chair: Laura Ventura**

C0272: **Statistical issues in cost-effectiveness analysis of medical treatments**
Presenter: **Elias Moreno**, University of Granada, Spain, Spain

A central problem in the Health Public System in most of the European countries is the selection of a treatment among a set of competitive treatments for a given disease. Since the resources are limited the selection of the treatment is to be based not only on their effectiveness but also on their cost. This problem is called the cost-effectiveness analysis of medical treatments, and it does not have along history although its interest is rapidly growing. We focus the cost-effectiveness analysis as a Statistical Decision Problem, and review the elements of the decision problem that include stochastic rewards and utility functions. For the main utility functions appearing in the literature optimal treatments are provided, and the methodology is illustrated on an example with real data. Unfortunately the data on cost and effectiveness of treatments are typically heterogenous, and this is a source of statistical difficulties. Some ideas about how to solve them are outlined.

C0307: **An alternative to objective Bayes model selection**
Presenter: **Monica Musio**, University of Cagliari, Italy
Co-authors: Philip Dawid

The use of "objective" improper prior distributions for Bayesian model selection is problematic, on account of the non-existence of a finite normalising constant. This leads to an undefined Bayes factor. There have been numerous attempts to circumvent this problem, all involving greater or lesser departure from coherent Bayesian principles. An alternative approach is presented, based on replacing the log score in the definition of the Bayes factor by a local proper scoring rule, such as that of Hyvarinen, which will not involve the normalising constant. The resulting model selection procedure, when conducted prequentially, will be consistent under mild conditions. The theory is illustrated by a number of examples.

C0257: **On Bayesian, confidence distribution and frequentist inference**
Presenter: **Erlis Ruli**, University of Padova, Italy
Co-authors: Laura Ventura

The aim is to discuss and characterize connections between frequentist, confidence distribution and objective Bayesian inference, when considering higher-order asymptotics, matching priors, and confidence distributions based on pivotal quantities. The focus is on testing precise or sharp null hypotheses on a scalar parameter of interest. Moreover, we illustrate that the application of these procedures requires little additional effort compared to the application of standard first-order theory. In this respect, using the **R** software, we indicate how to perform in practice the computation with two examples in the context of data from inter-laboratory and stress-strength reliability studies.

C0453: **Tools for robust Bayesian inference**
Presenter: **Laura Ventura**, University of Padova, Italy
Co-authors: Erlis Ruli, Nicola Sartori, Walter Racugno

The aim is to illustrate the properties and applications of the so-called robust pseudo-posterior distributions, i.e. posterior distributions derived from the combination of a pseudo-likelihood function or an unbised estimating function with suitable prior information. Examples of pseudo-likelihoods are the composite and the empirical likelihoods, while unbiased estimating functions include as special instances M-estimating functions and proper scoring rules.

CC080 Room Cocktail Hall STATISTICAL MODELLING **Chair: Vincent Vandewalle**

C0199: **Estimation of the fuzzy variance by different approximations of the fuzzy product**

Presenter: **Redina Berkachy**, ASAM, Applied Statistics And Modelling, Switzerland

Co-authors: Laurent Donze

Fuzzy statistical methods appear to be well suited to situations where the data we are collecting are exposed to fuzziness and uncertainty. Calculating for instance analytically or numerically the fuzzy variance could be advantageous. Yet, this task is not simple, especially regarding the difficulties in measuring the multiplication of two fuzzy sets. These computational problems are not evident to overcome. Therefore, an approximation of this product is needed. In the aim of computing the fuzzy variance, we propose different approximations of this product including the one using a particular method called the signed distance. However, using some of our approximations, another computational complexity arises since we get non-positive fuzzy numbers due to the difference between two fuzzy sets. This implies a result given by a fuzzy number violating the principles of the n-uples notations. In order to solve this problem, we use the shifting (translation) techniques. In addition, we give a comparison between these approximations in the purpose of displaying their characteristics. Finally, we illustrate our approach by numerical examples. We highlight that one should be prudent when choosing an estimation of the fuzzy variance.

C0406: **Inferring the parameters of arterial stiffness using gradient matching**

Presenter: **Benn Macdonald**, University of Glasgow, United Kingdom

Vessel stiffness affects pulmonary arterial pressure and flow and is a driver behind hypoxia (deficiency in the amount of oxygen reaching tissues). Statistically inferring the parameters of arterial stiffness usually involves numerically solving a set of partial differential equations (PDEs) for a given parameter set. The solution is compared to observations of pressure and flow and the likelihood or posterior density of the parameters can be calculated. This procedure is repeated as part of an iterative optimisation or sampling scheme (such as MCMC). However, the computational costs associated with repeatedly numerically solving the PDEs renders this method impractical for real-time decision-making in the clinic. Instead, gradient matching can be used to bypass solving the PDEs and therefore avoids the subsequent computational burden. This approach works by constructing a penalty or prior distribution that directly depends on the PDEs. All components of the equations are obtained from a smooth interpolant (in this case, a Gaussian process) and the parameters are then optimised by a penalised likelihood approach or sampled from the posterior distribution, subject to the soft-constraint that penalises deviations from the PDEs. This approach is applied to a fluid dynamics model in a simulation study to infer the parameters governing vessel stiffness.

C0196: **Bayesian and frequentist models for extrapolation of event rates for safety and efficacy data in clinical trials**

Presenter: **Daniel Bonzo**, LFB USA, United States

Co-authors: Marek Ancukiewicz

Event rates are one type of endpoint commonly used for evaluation of safety data (e.g., adverse events) and efficacy data (e.g., bleeding episodes). The data arise when one counts, for a patient, a random number of events observed during study follow-up of a random duration. We assume that event counts are overdispersed. We consider the problem of extrapolation of such data. Under certain circumstances, data can be extrapolated to a different population (e.g., from adult to pediatric population), a different but related indication, and different but similar product (e.g., from original drug to generic drug or biosimilar). As the concept of estimand captures population, endpoint, and a measure of effect in general, one can think about extrapolation of historical data from one estimand to another closely related estimand. We propose and evaluate two models for this task: random effect model using a non-parametric estimate of the variance of event rates (frequentist) and power model assuming partial exchangeability and negative-binomial likelihoods (Bayesian). We demonstrate application of the method using clinical trial data. In conclusion, both models require clinical and scientific inputs that can be quantified as weighting of different sources of data. The models can be useful tools for extrapolation of event rates data for medicine development, including extrapolation from adult to pediatric population, from one indication to another, and from one product to another.

C0361: **A censored generalized linear finite mixture model for the columbia card task**

Presenter: **Nienke Dijkstra**, Erasmus University Rotterdam, Netherlands

Co-authors: Patrick Groenen, Henning Tiemeier

The Columbia Card Task (CCT) is a card game that measures risk behavior. Participants gain points by turning over win cards, but turning a loss card ends the round and costs points. At any time the participant can stop playing a round. The aim is to estimate the number of cards a participant intends to turn over and relate this outcome to participant characteristics. When the participant faces a loss card, the observation is censored. The purpose is to build a statistical model that appropriately addresses the features of the CCT, which is performed in 16 rounds each by 3326 children aged 8-10 years. The new model should accommodate censoring and take into account the unobserved heterogeneity across individuals. We propose a censored generalized linear finite mixture model. Several distributions and link functions are investigated. They are compared based on their predictive performance, computational speed, convenience for usage, and their interpretablility. In particular, we study the Poisson distribution with an identity link that gives convenient linear effects. Alternatively, to weaken the equidispersion assumption of the Poisson distribution, we consider the negative binomial distribution.

CC086 Room Clio Hall NONPARAMETRIC METHODS **Chair: Hidetoshi Murakami**

C0377: **Statistical inference on stratified manifolds**

Presenter: **Charles Xi Yan**, Mr, United Kingdom

Stratified manifolds are metric spaces consisting of strata of different dimensions that are "stuck together" suitably. One type of stratified space is the so-called open book, the simplest of which is the 3-Spider. An important area of application of stratified spaces is to random tree structures, such as phylogenetic trees. However, it is not easy to construct parametric models on stratified manifolds. For this reason, a non-parametric approach to inference based on empirical likelihood has been investigated. By bootstrap replicates and bootstrap calibration, confidence regions can be constructed. To further the study, the core of such an inference, Wilk's Theorem, on different manifolds is introduced. Wilk's theorem has been proved by Owen on the Euclidean Space. Empirical likelihood is also applied to a second example, the unit sphere. By considering the extrinsic mean and intrinsic mean, Wilk's Theorem is proved again.

C0383: **Multiscale inference for nonaparametric time trends**

Presenter: **Marina Khismatullina**, University of Bonn, Germany

Co-authors: Michael Vogt

We develop multiscale methods to test qualitative hypotheses about nonparametric time trends. In many applications, practitioners are interested in whether the observed time series has a time trend at all, that is, whether the trend function is nonconstant. Moreover, they would like to get further information about the shape of the trend function. Among other things, they would like to know in which time regions there is an upward/downward movement in the trend. When multiple time series are observed, another important question is whether the observed time series all have the same time trend. We design multiscale tests to formally approach these questions. We derive asymptotic theory for the proposed tests and investigate their finite sample performance by means of simulations. In addition, we illustrate the methods by two applications to temperature data.

C0390: **A two-sample nonparametric statistic based on the power divergence**

Presenter: **Hidetoshi Murakami**, Tokyo University of Science, Japan

A testing hypothesis for the shifted location parameter in two-sample is considered. In two-sample problem, the Wilcoxon rank sum test is often used to test the location parameter. We propose a new nonparametric statistic based on the power divergence measure between the two distribution functions. In addition, we derive the exact critical value of suggested statistic for small sample sizes. Deriving the exact critical value of the statistic can be difficult when the sample sizes are increasing. Since an approximation method to the distribution function of the test statistic can be useful, we consider the moment-based approximation. Simulations are used to investigate the power of the proposed statistic with various distributions for small sample sizes. We compare the robustness to the actual significance level and the power of suggested statistic with various nonparametric statistics.

C0162: **Statistical inference in a spatial-temporal stochastic frontier model**

Presenter: **John Eustaquio**, University of the Philippines, Philippines

Co-authors: Erniel Barrios, Rouselle Lavado

The stochastic frontier model with heterogeneous technical efficiency explained by exogenous variables is augmented with a spatial-temporal component, a generalization relaxing the panel independence assumption in a panel data. The estimation procedure takes advantage of additivity in the model, computational advantages over simultaneous maximum likelihood estimation of parameters is exhibited. Estimates of technical efficiency estimates are comparable to existing models estimated with maximum likelihood methods. The spatial-temporal component can improve estimates of technical efficiency in a production frontier that is usually biased downwards. We present a test to verify model assumptions that facilitates estimation of parameters.

CC087 Room Vega Hall MACHINE LEARNING AND DATA SCIENCE	**Chair: Emmanuele Sordini**

C0347: **Moment distances for comparing high-entropy distributions with application in domain adaptation**

Presenter: **Werner Zellinger**, Johannes Kepler University Linz, Austria

Co-authors: Bernhard Moser, Hamid Eghbal-zadeh, Michael Zwick, Edwin Lughofer, Thomas Natschlaeger, Susanne Saminger-Platz

Given two samples, the similarity of the distributions of the sample representations in the latent space of a discriminative model shall be enforced. Standard approaches are based on the minimization through probability metrics, e.g. by the Wasserstein metric, the Maximum Mean Discrepancy, or f-divergences. However, also moment distances not satisfying the identity of indiscernibles, i.e. pseudo-metrics, performed well in many practical tasks. The uniform distance between two distributions having finitely many moments in common can be very large. The question is under which constraints on the distributions small values of moment distances imply distribution similarity. We show that the total variation distance between two distributions is small if the distributions are of high differential entropy constrained at finitely many moments that are similar for the two distribution. We also discuss existing relations between moment convergence and moment-constrained entropy convergence in the one-dimensional case. Our analysis leads to a new target error bound for domain adaptation, which is underpinned by numerical evaluations.

C0403: **Exploratory data analysis about the optimal number of hidden layers and nodes in deep neural network**

Presenter: **Jae Eun Lee**, Pukyong National University, Korea, South

Co-authors: Dae-Heung Jang

To find out the optimal network configuration about the number of hidden layers and nodes in deep neural network, we need to use the exploratory data analysis (EDA) approach. This is because it is difficult to discover theoretical approach. We perform EDA for finding the optimal network configuration about the number of hidden layers and nodes in deep neural network through some examples of the prediction and the classification. When we try the prediction and the classification about datasets using deep neural network, we examine whether the deep neural network can decide the optimal number of hidden layers and nodes through the EDA approach.

C0420: **100+ years of graphs of the Titanic data**

Presenter: **Juergen Symanzik**, Utah State University, United States

Co-authors: Michael Friendly, Ortac Onder

Many readers are likely familiar with the stories of the tragic fate of passengers and crew of the RMS Titanic upon her fatal collision with an iceberg and her sinking in the early hours of April 15, 1912 on her maiden voyage to New York City. Little known is the fact that the first graphical summary of the initial survivor data appeared in The Sphere, a British newspaper, on May 4, 1912. The public inquiries that followed produced detailed data sets that have been widely used to illustrate graphical and statistical methods for quite some time. Numerous follow-up studies have used a wide variety of graphical representations related to the Titanic disaster, published in statistics, information visualization, and social sciences venues. It seemed timely to survey the variety of graphical methods used for these data sets over the last century. Graph types used to portray the Titanic data include: bar charts, mosaic plots, double-decker plots, parallel set plots, Venn diagrams, balloon plots, nomograms, and tree diagrams to name only a few. Three questions are addressed: (i) What types of graphs have been used for the Titanic data in the last 100+ years? (ii) Are some of these graphs unique and provide additional insights that are hard to obtain from the other ones? (iii) Do the graphs used differ by genre or scientific discipline?

C0401: **How to teach a machine to read an electrocardiograph**

Presenter: **Thomas Alexander Gerds**, University of Copenhagen, Denmark

Co-authors: Andreas Kryger Jensen

The statistical challenge to train a computer for the task of interpreting a digital electrocardiograph (ECG) is considered. The aim is to predict the current health state of the patient's heart and the development of future heart diseases. To capture the signal in these highly correlated data we develop a two-step approach. The first step consists of a series of unsupervised functional data analyses performed separately for each of the 12 leads of the ECG. The second step uses the health outcome to combine the lead specific patterns into a personalized prediction. For each outcome separately, we train the model regarding the sparse multivariate principal modes of association in order to achieve optimal predictive performance invalidation data.

CG027 Room Orion Hall ADVANCES IN MULTIVARIATE ANALYSIS AND BIG DATA	**Chair: Florian Frommlet**

C0332: **An outlier detection method for high dimensional data with mixed attributes**

Presenter: **Kangmo Jung**, Kunsan National University, Korea, South

Outliers are extreme observations which are far away from other observations. Outlier detection becomes a significant procedure for many applications such as detecting insurance fraud or industrial damage. Most of the research work in outlier detection has focused on data sets having one type of attribute, that is, only continuous attributes or categorical attributes. Furthermore, in these days the data sets with many attributes tend to be sparse, and conventional methods using the Euclidean distance or nearest neighbors become inappropriate. We propose an outlier detection method using both quantiles of attribute value frequency score for categorical attributes and the Mahalanobis distance for continuous attributes. It also handles sparse high-dimensional continuous data and it is very fast, scalable. Experimental results show how the proposed method compares with other state-of-the art outlier detection methods proposed in the literature.

C0304: **Fast simulation-based estimation for complex models**

Presenter: **Maria-Pia Victoria-Feser**, University of Geneva, Switzerland

Co-authors: Stephane Guerrier, Samuel Orso

Along the ever-increasing data size and model complexity, an important challenge frequently encountered in constructing new estimators or in implementing a classical one such as the maximum likelihood estimator, is the computational aspect of the estimation procedure. To carry out estimation, approximate methods such as pseudo-likelihood functions or approximated estimating equations are increasingly used in practice as these methods are typically easier to implement numerically although they can lead to inconsistent and/or biased estimators. In this context, we extend and provide refinements on the known bias correction properties of two simulation-based methods, respectively indirect inference and bootstrap, each with two alternatives. These results allow one to build a framework defining simulation-based estimators that can be implemented for complex models. Indeed, as previously shown, based on a biased or even inconsistent estimator, several simulation-based methods can be used to define new estimators that are both consistent and numerically very fast to compute in complex settings. This framework includes the classical method of indirect inference without requiring specification of an auxiliary model. We illustrate the use of simulation-based estimation, with initial estimators that are fast to compute, in the framework of Generalized Linear Models and (exploratory) factor analysis with binary outcomes when $p > n$, with large p.

C0341: **Narrow big data in streams and Kolmogorov complexity**

Presenter: **Michal Cerny**, University of Economics Prague, Czech Republic

The following computational model for Narrow Big Data is considered. The dataset is formalized as an $(n \times p)$-matrix A where n stands for the number of observations, p stands for the dimension and n is assumed to be superpolynomial in p. The memory is restricted by a polynomial in p. Thus, A cannot be stored in memory in full. Rows of A (data points) are accessible on-line one-by-one; once a data point is read into memory, it is dropped forever. This is a natural model for representation of Narrow Big Data which, however, imposes serious restrictions on statistical algorithms. It can be shown that certain quantities cannot be computed in the model at all. As an example, we prove that the sample median cannot be computed in this model. Negative proofs are based on Kolmogorov complexity arguments, and essentially show that such quantities contain "too much information" which cannot be stored within the memory bounds imposed by the computational model. Another example is linear regression - while Ordinary Least Squares (OLS) can be efficiently computed in the model by a sequence of rank-one updates of the information matrix, Kolmogorov complexity arguments imply that L_1-norm estimators cannot be computed at all.

C0373: **A dynamic approach for clustering of variables in high-dimensional analysis**

Presenter: **Christian Derquenne**, EDF Research and Development, France

The research of structures in the data represents an essential aid to understanding the phenomena to be analyzed. We have previously offered a set of methods for clustering numeric variables with linear or non-linear links. In case of high-dimensional data (a lot of variables and a lot of individuals), we propose to adapt these methods by means of different strategies to divide to conquer. Firstly, a random sample of individuals is collected and it is cut in random groups of variables. The sample of individuals and the groups of variables have reasonable sizes. On each group, a clustering method is applied. Each cluster is represented by the first principal component. Then, the same clustering method is applied again on all first principal components, and a final typology is obtained grouping initial variables. However, this process has only been applied on one random sample of individuals. So we evaluate the quality of the results, and we extend them with different strategies depending on such quality. Finally, a multiple correspondence analysis is applied to obtain a last typology with all data. This approach has been applied to simulated data and real data.

Friday 31.08.2018 09:00 - 10:30 Parallel Session L – COMPSTAT2018

CI004 Room Cuza Hall COMPUTATIONAL ECONOMETRICS Chair: Erricos John Kontoghiorghes

C0325: On the asymmetric impact of macrovariables on volatility

Presenter: **Alessandra Amendola**, Department of Economics and Statistics - University of Salerno, Italy
Co-authors: Vincenzo Candila, Giampiero Maria Gallo

The GARCH-MIDAS is extended to take into account possible different impacts from positive and negative macroeconomic variations on financial market volatility. We evaluate the proposed specification by a Monte Carlo simulation which shows good estimation properties with the increase in the sample size. The empirical application is performed on the daily S&P500 volatility dynamics with the U.S. monthly industrial production and national activity index as additional (signed) determinants. In the out-of-sample analysis, our proposed GARCH-MIDAS model statistically outperforms the competing specifications, represented by different symmetric and asymmetric GARCH model specifications.

C0436: Filters, waves and spectra

Presenter: **Stephen Pollock**, University of Leicester, United Kingdom

Econometric analysis requires filtering techniques that are adapted to cater to data sequences that are short and that have strong trends. Whereas the economists have tended to conduct their analyses in the time domain, the engineers have emphasised the frequency domain. Emphasis is placed in the frequency domain; and it is shown how the frequency-domain methods can be adapted to cater to short trended sequences. Working in the frequency domain allows an unrestricted choice to be made of the frequency response of a filter. It also requires that the data should be free of trends. Methods for extracting the trends prior to filtering and for restoring them thereafter are described.

C0220: Comparing trends in topics in economic journals overtime

Presenter: **Peter Winker**, University of Giessen, Germany
Co-authors: David Lenz

The comparison of information content of different sources is highly relevant. We compare text corpora, specifically articles published in two economic journals. Thereby, the focus is on the development of topic importance over time and how it correlates across journals. Similar questions arise in many fields of practical relevance and, if at all, are mostly answered impressionistically. We present a quantitative framework for comparing text corpora based on their latent topics using text mining techniques. Paragraph Vector Topic Modeling is applied to identify latent topics in text corpora and time information is utilized to track the evolution of these topics. This allows the comparison of corpus compositions over time. More specifically, we evaluate three comparison methods: Treat both text corpora as a single corpus, train a model on one corpus and evaluate the other corpus based on this model and vice versa, and train a model for each corpus and use a matching approach for pairing corresponding topics. For the empirical application, we exploit the corpus of articles published in the Journal of Economics and Statistics and the corpus of articles published in the Review of World Economics, both from 1913 to 1940. We present topic dynamics for both corpora and information on how strong the correlation of these dynamics was across journals. Furthermore, the analysis indicates which of the methods presented above is most promising for this type of analysis.

CO070 Room Mezzanine Lounge FLEXIBLE MODELING Chair: Francesco Lagona

C0155: A tractable multi-partitions clustering

Presenter: **Vincent Vandewalle**, Inria, France
Co-authors: Matthieu Marbac-Lourdelle

In the framework of model-based clustering, a model allowing several latent class variables is proposed. This model assumes that the distribution of the observed data can be factorized into several independent blocks of variables. Each block is assumed to follow a latent class model (i.e., mixture with conditional independence assumption). The proposed model includes variable selection, as a special case, and is able to cope with the mixed-data setting. The simplicity of the model allows to estimate the repartition of the variables into blocks and the mixture parameters simultaneously, thus avoiding running EM algorithms for each possible repartition of variables into blocks. For the proposed method, a model is defined by the number of blocks, the number of clusters inside each block and the repartition of variables into block. Model selection can be done with two information criteria, the BIC and the MICL, for which an efficient optimization is proposed. The performances of the model will be studied on simulated and real data. It will be shown that the proposed method gives a rich interpretation of the dataset at hand (i.e., analysis of the repartition of the variables into blocks and analysis of the clusters produced by each block of variables).

C0244: A non-homogeneous hidden Markov model for partially observed longitudinal responses

Presenter: **Maria Francesca Marino**, University of Florence, Italy
Co-authors: Marco Alfo

Dropout represents a typical issue in longitudinal studies. If the mechanism that generates the missing data is non-ignorable, inference based only on the observed data may be severely biased. Therefore, it is worth to define a model that describes the dropout process and links this auxiliary model to the main one, entailing the longitudinal responses. A frequent strategy is based on using individual-specific random coefficients that help capture sources of unobserved heterogeneity and introduce a reasonable structure of dependence between the longitudinal and the missing data process. In this way, we model dependence within and between profiles from the same subject using two different, but correlated, sets of random coefficients. For the longitudinal outcome, we consider time-varying (discrete) random coefficients that evolve over time according to a non-homogeneous hidden Markov chain. The aim is to capture differential (possibly non-smooth) dynamics in the individual longitudinal profiles. For the missing data indicator, we consider time-constant (discrete) random coefficients to represent the effect of the individual-specific propensity to stay into the study. Dependence between profiles is described by an upper-level latent class variable that allows to connect the two sets of random coefficients.

C0374: Gradient boosting in generalized Markov-switching regression models

Presenter: **Timo Adam**, Bielefeld University, Germany
Co-authors: Andreas Mayr, Thomas Kneib, Roland Langrock

Markov-switching generalized additive models for location, scale and shape constitute a novel class of latent-state time series regression models that allow different state-dependent parameters of the response distribution - not only the mean, but also variance, skewness and kurtosis parameters - to be modelled as potentially smooth functions of a given set of explanatory variables. In addition, the set of possible distributions that can be specified for the response is not limited to the exponential family but additionally includes, for instance, a variety of Box-Cox-transformed, zero-inflated and mixture distributions. An estimation approach based on the EM algorithm is proposed, where the gradient boosting framework is used to prevent overfitting while simultaneously performing variable selection. The feasibility of the suggested approach is assessed in simulation experiments and illustrated in a real-data setting, where the conditional distribution of the daily average price of energy in Spain is modeled over time.

C0345: Supervised multivariate discretization and levels merging for logistic regression

Presenter: **Adrien Ehrhardt**, Inria, France

Co-authors: Vincent Vandewalle, Christophe Biernacki, Philippe Heinrich

For regulatory and interpretability reasons, the logistic regression is still widely used by financial institutions to learn the refunding probability of a loan given the applicants characteristics from historical data. Although logistic regression handles naturally both quantitative and qualitative data, three ad hoc pre-processing steps are usually performed: firstly, continuous features are discretized by assigning factor levels to pre-determined intervals; secondly, qualitative features, if they take numerous values, are grouped; thirdly, interactions (products between two different features) are sparsely introduced. By reinterpreting these discretized (resp. grouped) features as latent variables and by modeling the conditional distribution of each of these latent variables given each original feature with a polytomous logistic link (resp. contingency table), a novel model-based resolution of the discretization problem is introduced. Estimation is performed via a Stochastic Expectation-Maximization (SEM) algorithm and a Gibbs sampler to find the best discretization (resp. grouping) scheme w.r.t. any classical logistic regression loss (AIC, BIC, test set AUC, ...). For detecting interacting features, the same scheme is used by replacing the Gibbs sampler by a Metropolis-Hastings algorithm. The good performances of this approach are illustrated on simulated and real data from Credit Agricole Consumer Finance.

CO092 **Room Vega Hall** TUTORIAL 1 **Chair: Frederic Ferraty**

C0450: **Functional data and nonparametric modelling: Theoretical/methodological/practical aspects**

Presenter: **Frederic Ferraty**, Mathematics Institute of Toulouse, France

Situations when one observes a response (scalar or functional variable) and functional predictor(s) are considered. The natural statistical question is very simple: are we able to predict correctly the response from the functional predictor(s) when one has no idea on the relationship between the response and functional predictor(s)? A suitable answer to this important statistical issue is the "functional nonparametric regression". The word "nonparametric" stands for any model requiring very few assumptions with respect to the relationship between the response and the predictor(s); the word "functional" reminds that the model has to handle functional data. So, the aim is to give an extensive overview on this statistical topic. In addition to some theoretical and practical key developments, real datasets illustrate the purpose (benchmark datasets, hyperspectral image, forensic entomology in the context of criminology, etc).

CC076 **Room Clio Hall** ROBUST STATISTICS AND HEAVY TAILS **Chair: Aldo Corbellini**

C0407: **Robust estimation and variable selection for regression models using empirical likelihood and LASSO methods**

Presenter: **Senay Ozdemir**, Afyon Kocatepe University, Turkey

Co-authors: Yesim Guney, Yetkin Tuac, Olcay Arslan

There are several methods to estimate the parameters of a linear regression model. These methods are mainly based on some distributional assumptions on error terms. When these assumptions are not plausible, nonparametric methods, like empirical likelihood, can be used to deal with the estimation problem in regression analysis. Empirical likelihood method performs by maximizing an empirical likelihood function defined as the multiplication of unknown probabilistic weights for each observation under some constraints. In ordinary case, one of these constraints is similar to normal equation in ordinary least square estimation method which is drastically effected from outliers in the data. Replacing this non-robust constraint with a robust one, the empirical likelihood estimators can be made robust against the outliers in data. Another vital issue in a regression analysis is to select the significant variables. It has been mainly purposed to also carry on variable selection along with the parameter estimation in a regression analysis based on the empirical likelihood method. To this extend, we combine the LASSO (least absolute shrinkage and selection operator) method with robust empirical likelihood regression estimation to obtain robust regression estimators and select the important explanatory variables.

C0410: **Robust estimation and variable selection in joint location and scale model using least favorable distributions**

Presenter: **Yesim Guney**, Ankara University, Turkey

Co-authors: Yetkin Tuac, Senay Ozdemir, Olcay Arslan

The assumption of equal variances is not always appropriate and different approaches to modelling variance heterogeneity have been widely studied in the literature. One of these approaches is joint location and scale model (JLSM) defined with the idea that both the location and the scale depend on explanatory variables through parametric linear models. Because JLSM includes two models in itself, it does not deal well with many irrelevant variables. Therefore, determining the variables affecting the location and the scale is as important as estimating the parameters of this model. From this point of view, a combine robust estimation and variable selection method is proposed to simultaneously estimate the parameters and select the important variables. This is done using the least favorable distribution and LASSO method.

C0434: **Robust principal volatility components**

Presenter: **Carlos Trucios**, Sao Paulo School of Economics - FGV, Brazil

The recent principal volatility components procedure is analized. The procedure overcomes several difficulties in modelling and forecasting the conditional covariance matrix in large dimensions arising from the curse of dimensionality. We take into account the presence of outliers, which are common in financial time series, and show that outliers have a devastating effect on the construction of the principal volatility components and on the forecast conditional covariance matrix and consequently in economic and financial applications based of this forecast. To overcome this problem, we propose a robust procedure called robust principal volatility components and analyse its finite sample properties by means of Monte Carlo experiments. The procedure is also illustrated using empirical data. The robust procedure outperforms the non robust method in simulated and empirical data.

C0417: **Vector autoregressive models with multivariate skew innovations**

Presenter: **Yetkin Tuac**, Ankara University, Turkey

Co-authors: Yesim Guney, Senay Ozdemir, Olcay Arslan

Multiple time series analysis can be done with the help of vector autoregressive (VAR) models and the parameter estimation of these models are usually done under normality assumption. Since normality assumption is too restrictive for real data analysis, some heavy tailed alternatives, such as the t distribution, have been proposed in literature as the distribution for innovations. However, real data examples show that if the skewness is present, then the parameter estimation from normal or heavy tailed symmetric distributions will be produced inefficient estimators for the parameters of a VAR model. Therefore, if skewness and/or heavy taildness is a concern asymmetric and/or heavy tailed distributed innovation of a VAR model should be considered as alternatives to the symmetric distributed innovations.

CG009 **Room Cocktail Hall** ADVANCES IN SEMI- AND NON-PARAMETRIC MODELING **Chair: Ivette Gomes**

C0177: **Unit root test in a semiparametric model**

Presenter: **Sarah Bernadette Aracid**, University of the Philippines Los Banos, Philippines

Co-authors: Erniel Barrios, Joseph Ryan Lansangan

Presence of unit root in time series data is implicated in the persistent effect of random shocks in the behavior of a model, leading most unit root tests to be incorrectly-sized or have low power or both. A nonparametric test for the presence of unit root is proposed. To better understand the instance where unit root occurs, hence, mitigate the possible problem of present unit root tests, it is assumed that another time series x_t possibly affect the target time series y_t in addition to the autocorrelation dynamics. A nonparametric effect of x_t can spare the autocorrelation structure from further contaminations, hence, the test can characterize presence of unit roots in y_t easily. Simulation study showed that the proposed test yields better size and power compared to some popular tests for unit root.

C0182: **A nonparametric test for intervention effect in Markov regime switching model**

Presenter: **Mara Sherlin Talento**, College of Arts and Sciences, University of the Philippines Los Banos, Philippines

Co-authors: Erniel Barrios, Joseph Ryan Lansangan

Highly volatile time series data like those in the financial markets usually exhibit regime switching process whose behavior becomes more complicated when interventions like policy change or contagion induces effects on the model dynamics. We developed a method of testing presence of intervention effect in a Markov regime switching model. The test allows the regimes to be affected by varying nature of interventions at different time points. A simulation study is designed to account for example different combination of intervention types, magnitude of intervention effect, and distance of two regimes. The simulation study exhibits that the test is correctly sized and powerful especially when the intervention effect is postulated to occur with a pulse-type intervention variable.

C0184: **Nonparametric test for cointegration**

Presenter: **Jan Andrew Reforsado**, University of the Philippines Los Banos, Philippines

Co-authors: Erniel Barrios, Joseph Ryan Lansangan

Cointegration testing is an important aspect of modeling in nonstationary time series data to avoid the possibility of observing spurious relationship among variables. Existing tests for cointegration usually exhibit less optimal behavior specially for short time series data. A vector error correction (VEC) model is estimated through the backfitting algorithm, the fitted model is used in replicating the data through sieve bootstrap. The empirical distribution of eigenvalues from the lagged error correction matrix generated from the data replicates are used in testing for cointegration. Simulation study shows that the proposed nonparametric test yields size and power at least comparable to some well-known tests for cointegration.

C0175: **Semiparametric mixed model for analysis of covariance with high dimensional covariates**

Presenter: **Erniel Barrios**, University of the Philippines, Philippines

Co-authors: Stephen Jun Villejo, Joseph Ryan Lansangan

Treatment effects are difficult to measure from clinical trials involving live human subjects due to the contamination of the response resulting that non-homogeneous experimental units. The experiment should be designed to allow measurement of as many covariates as possible, so that responses are adjusted to facilitate estimation of treatment effects. There could be more covariates than the number of experimental units that manifest the responses, leading to a potentially overparameterized model and prone to observing false positive evidence on treatment effect. We consider an additive mixed semiparametric model to adjust the responses due to heterogeneity of experimental units indexed by the covariates. We then simultaneously estimate the treatment effect while simultaneously reducing the dimension of the covariates. Simulation studies are conducted, the methods are also used in actual clinical trials.

C0181: **Nonparametric changepoint analysis in multiple time series**

Presenter: **Elfred John Abacan**, College of Arts and Sciences, University of the Philippines Visayas, Philippines

Co-authors: Erniel Barrios, Joseph Ryan Lansangan

Analysis on changes in the level of time series helps in characterizing common components of multiple time series that helps identify the shared behavior of the data generating process with some known events that causes perturbations in the behavior of the time series. Change in variance of the error structure leads to model misspecification and more serious violation of other assumptions that facilitates model estimation. Volatility models are used to incorporate variance structure into the mean model, but sometimes, this suffers from overparameterization, especially in multiple series data. A model for structural change in the variance component is estimated through the backfitting algorithm, this is used then as a reference for a test based on sieve bootstrap to detect changes in the variance of multiple time series. A simulation study shows the test performs well in terms of power and size.

Friday 31.08.2018 11:00 - 12:00 Parallel Session M – COMPSTAT2018

CI016 Room Cuza Hall ADVANCES IN FUNCTIONAL DATA ANALYSIS Chair: Fang Yao

C0370: **Inference in separable Hilbert spaces using Hotelling's T2**
Presenter: **Aymeric Stamm**, Human Technopole - IIT, Italy
Co-authors: Alessia Pini, Simone Vantini

Hotelling's T2 is introduced in multivariate data analysis courses for parametric hypothesis testing on the mean vector of multivariate Gaussian distributions. In details, given a sample of n i.i.d. random variables following a p-variate Gaussian distribution, under the null hypothesis that the mean vector is equal to some fixed value, a properly scaled version of Hotelling's T2 statistic follows a Fisher distribution, provided that $n > p$. When either the data does not follow a Gaussian distribution or its dimension exceeds the sample size, this result does not hold anymore, which has led the statistical community to move away from Hotelling's T2 and introduce new statistics along with non-parametric approaches to solve one- and two-sample testing problems. Situations like this naturally arise from high-dimensional data (i.e. when $p > n$), from functional data (where p is actually infinite) or, more generally, from object data which belong to Hilbert spaces or even metric spaces. We will show that Hotelling's T2 is in fact well defined in generic Hilbert spaces, and we will provide a practical way of computing its value in separable Hilbert spaces. Next, we will provide an exact permutation testing procedure based on Hotelling's T2 statistic for solving the one- and two-sample problems in separable Hilbert spaces. We will show simulations and a case study on Aneurysm data as basis for discussing the performances of the approach.

C0447: **Functional regression on manifolds with contamination**
Presenter: **Fang Yao**, Peking University, University of Toronto, Canada
Co-authors: Zhenhua Lin

The focus is on a new perspective on functional regression with a predictor process via the concept of manifold that is intrinsically finite-dimensional and embedded in an infinite-dimensional functional space, where the predictor is contaminated with discrete/noisy measurements. By a method of functional local linear manifold smoothing, we achieve a polynomial rate of convergence that adapts to the intrinsic manifold dimension and the level of sampling/noise contamination with a phase transition phenomenon depending on their interplay. This is in contrast to the logarithmic convergence rate in the literature of functional nonparametric regression. We demonstrate that the proposed method enjoys favorable finite sample performance relative to commonly used methods via simulated and real data examples.

C0455: **Dynamic modelling with Data2PDE**
Presenter: **Michelle Carey**, Univerity College Dublin, Ireland
Co-authors: James Ramsay

Geo spatial data are observations of a process that are collected in conjunction with reference to their geographical location. This type of data is abundant in many scientific fields, some examples include: population census, social and demographic, economic and environmental data. They are often distributed over irregularly shaped spatial domains with complex boundaries and interior holes. Modelling approaches must account for the spatial dependence over these irregular domains as well as describing the temporal evolution. Dynamic systems modelling has a huge potential in statistics, as evidenced by the amount of activity in functional data analysis. Many seemingly complex forms of functional variation can be more simply represented as a set of differential equations, either ordinary or partial. We will present a class of semi parametric regression models with differential regularization in the form of PDEs. This methodology will be called Data2PDE Data to Partial Differential EquationsḊata2PDE characterizes spatial processes that evolve over complex geometries in the presence of uncertain, incomplete and often noisy observations and prior knowledge regarding the physical principles of the process characterized by a PDE.

CO034 Room Cocktail Hall DIRECTIONAL STATISTICS Chair: Francesco Lagona

C0164: **Dividing the Iberian peninsula using wildfires data**
Presenter: **Jose Ameijeiras-Alonso**, University of Santiago de Compostela, Spain
Co-authors: Francesco Lagona, Monia Ranalli, Rosa Crujeiras

The aim is to present a model for providing a spatial segmentation of wildfire occurrences in the Iberian Peninsula according to a finite number of latent classes employing a hidden Markov random field. A model based on a mixture of Kato-Jones circular densities is suggested. This model takes into account special features of wildfire occurrence data such as multimodality, skewness and kurtosis. The parameters of the model vary across space according to a latent Potts model, modulated by geo-referenced covariates. Also, due to the numerical intractability of the likelihood method when estimating these parameters, a composite-likelihood method will be presented for solving this issue.

C0222: **The joint projected normal and skew-normal: A distribution for poly-cylindrical data**
Presenter: **Gianluca Mastrantonio**, Politecnico of Turin, Italy

The aim is to introduce a multivariate circular-linear (or poly-cylindrical) distribution obtained by combining the projected and the skew-normal. We show the flexibility of our proposal, its closure under marginalization, and how to quantify multivariate dependence. Due to a non-identifiability issue that our proposal inherits from the projected normal, a computational problem arises. We overcome it in a Bayesian framework, adding suitable latent variables and showing that posterior samples can be obtained with a post-processing of the estimation algorithm output. Under specific prior choices, this approach enables us to implement a Markov chain Monte Carlo algorithm relying only on Gibbs steps, where the updates of the parameters are done as if we were working with a multivariate normal likelihood. The proposed approach can also be used with the projected normal. The proposal is used in a real data example, where the turning-angles (circular variables) and the logarithm of the step-lengths (linear variables) of four dogs are modeled jointly.

C0237: **Angle-based models for ranking data**
Presenter: **Philip Yu**, The University of Hong Kong, Hong Kong
Co-authors: Mayer Alvo, Hang Xu

A new class of general exponential ranking models is introduced which we label angle-based models for ranking data. A consensus score vector is assumed, which assigns scores to a set of items, where the scores reflect a consensus view of the relative preference of the items. The probability of observing a ranking is modeled to be proportional to its cosine of the angle from the consensus vector. Bayesian variational inference is employed to determine the corresponding predictive density. It can be seen from simulation experiments that the Bayesian variational inference approach not only has great computational advantage compared to the traditional MCMC, but also avoids the problem of overfitting inherent when using maximum likelihood methods. The model also works when a large number of items are ranked which is usually an NP-hard problem to find the estimate of parameters for other classes of ranking models. Model extensions to incomplete rankings and mixture models are also developed. Real data applications demonstrate that the model and extensions can handle different tasks for the analysis of ranking data.

CO105 Room Clio Hall LARS-IASC SESSION: RECENT ADVANCES IN STATISTICAL COMPUTING Chair: Paulo Canas Rodrigues

C0201: **A robust approach to singular spectrum analysis**
Presenter: **Paulo Canas Rodrigues**, Federal University of Bahia, Brazil

Singular spectrum analysis (SSA) is a non-parametric method for time series analysis and forecasting that incorporates elements of classical time series analysis, multivariate statistics, multivariate geometry, dynamical systems and signal processing. Although this technique has shown to be advantageous over traditional model based methods, in particular, one of the steps of the SSA algorithm, which refers to the singular value decomposition (SVD) of the trajectory matrix, is highly sensitive to data contamination. Specifically, and because SVD decomposition is least-squares based, the presence of a single outlier, if extreme, may be enough to draw the leading principal component towards itself resulting in possible misinterpretations, which may subsequently, in particular and in the case of time series analysis, lead to reduced quality of model fit and forecast accuracy. In order to alleviate this problem, a robust SSA algorithm is proposed, where a robust SVD procedure replaces the least-squares based one in the original SSA procedure. The SSA and the robust SSA approaches are compared in terms of model fit quality via Monte Carlo simulations that contemplate both clean and noisy/contaminated time series. The adequacy and value of the proposed approach is then compared with the standard SSA using real data.

C0280: **Efficient closed-form MAP estimators for some survival distributions**
Presenter: **Francisco Louzada**, University of Sao Paulo, Brazil

Nakagami-type distributions play an important role in communication engineering problems, particularly to model fading of radio signals. A maximum a posteriori (MAP) estimator for the Nakagami-type fading parameter is proposed. The MAP estimator has a simple closed-form expression and can be rewritten as a bias-corrected generalized moment estimator. Numerical results demonstrate that the MAP estimation scheme outperforms the existing estimation procedures and produces almost unbiased estimates for the fading parameter even for small sample size. In a second stage of the presentation, we show that the obtained results can be extended to other probability distributions, such as the gamma and generalized gamma ones. The potentiality of our proposed methodology is illustrated in a real reliability data set.

C0441: **Prediction and optimal sampling for spatial multivariate functional random fields**
Presenter: **Martha Bohorquez**, Universidad Nacional de Colombia, Colombia
Co-authors: Ramon Giraldo, Jorge Mateu

The framework of optimal sampling designs is extended to the spatial prediction of univariate and multivariate functional data. In both cases, we derive unbiased predictors and their variances. In the univariate case, we propose to use a simple cokriging predictor with the scalar random fields resulting from the scores associated to the representation of the functional data with the empirical functional principal components. In the multivariate case, we develop spatial prediction of a functional variable at unsampled sites, using functional covariates; that is, we present a functional cokriging method. We show that through the representation of each function in terms of its empirical functional principal components, the functional cokriging only depends on the auto-covariance and cross-covariance of the associated score vectors, which are scalar random fields. Design criteria are given for all predictors derived in this thesis. The methodologies are applied to the networks of air quality of Bogota and Mexico.

CO052 **Room Mezzanine Lounge** **ROBUST STATISTICS AND DATA SCIENCE WITH R** **Chair: Valentin Todorov**

C0261: **fsdaR: Making the FSDA toolbox available to R users**
Presenter: **Emmanuele Sordini**, Joint Research Centre of the European Commission (JRC), Italy
Co-authors: Valentin Todorov, Aldo Corbellini

The FSDA (Flexible Statistics for Data Analysis) toolbox is a software library that extends MATLAB and its Statistics Toolbox to support a robust and efficient analysis of complex datasets, affected by different sources of heterogeneity. FSDA was born around the forward search approach, and has evolved into a comprehensive and computationally efficient software package for robust statistics. FSDA provides tools for regression and multivariate analysis, robust transformations, cluster analysis and model selection. FSDA provides a rich set of graphical features not available in R, such as dynamic brushing, linking, etc., particularly useful for exploratory data analysis. The only downside is that FSDA is based on commercial software (MATLAB), which is not so appealing to the majority of the statistical community, where R is more widespread. An R package (called fsdaR) compliant with R coding and calling standards (object-oriented approach, formula notation, etc.) is now available in CRAN. This package provides to R users selected features of the FSDA toolbox along with some of its advanced graphical capabilities. Among these features are regression and multivariate analysis with all the diagnostic and exploratory data analysis features available in FSDA. Our presentation will focus on i) The structure of the fsdaR package; ii) computational and graphical features of FSDA available in R through fsdaR, iii) A live demo.

C0301: **Robust methods for ranking using composite indicators**
Presenter: **Wenjing Wang**, Vrije Universiteit Brussel, Belgium
Co-authors: Kris Boudt, Valentin Todorov

Composite indicators (CIs) are widely used for evaluating and ranking multidimensional individual performance, such as countries industrial competitiveness or well-being. Their composite nature necessitates standardizing the individual indicators prior to the aggregating. Popular choices include the use of so-called z-scores and min-max standardization. The former is popular in combination with arithmetic aggregation, while the latter is often used in conjunction with geometric aggregation. Both the z-score and min-max standardization approaches have in common that, because of the standardization, the presence of an extreme observation in the time series leads to an explosion of the scale statistics (standard deviation, range) and thus an implosion to zero of most observations in the standardized time series of that indicator. From the viewpoint of ranking using individual indicators, this has of course no effect on the ranking, but, as we show, it can have large effects on the ranking obtained using the composite indicator. We document in detail the robustness issues of standard approaches to ranking using composite indicators, and propose alternatives that are less sensitive to extreme observations in the data. The main contribution is that we propose a framework to use distribution-based winsorization methods to reduce the impact of outliers on ranking using composite indicators.

C0330: **R scripts for automated K-means clustering**
Presenter: **Sung-Soo Kim**, Korea National Open University, Korea, South

The aim is to provide R scripts for an automated K-means clustering process which decide the number of clusters and the initial cluster center, which is combined with variable selection process in a forward manner. We also provide potential outliers whenever a new variable is added in a K-means clustering process. This process can be used to see the inter-relationship between variables and outliers, and interactive data visualization is combined to clarify the clustering structure and help to identify outliers.

CO030 **Room Vega Hall** **RECENT ADVANCES IN COPULA-BASED MODELS** **Chair: Elisa Perrone**

C0310: **Dependence properties of l-infinity spherical densities and related models**
Presenter: **Florentina Suter**, University of Bucharest and "Gheorghe Mihoc-Caius Iacob', Romania

Spherical densities in the l-infinity norm are exchangeable densities with properties that are of interest in fields like statistical analysis of life data, or the reliability theory. Such densities functions are easy to identify and construct, and one of their property is that the n-dimensional density is the density of the n first epoch times of a nonhomogeneous pure birth process. They share this property with sequential order statistics and with load-sharing models from the reliability theory. The aim is to explore the dependence structure of l-infinity spherical and related models in terms of the multivariate copula, as well as some connections and applications that result from it.

C0285: **Data-driven regression and uncertainty quantification by polynomial chaos expansions and copulas**
Presenter: **Emiliano Torre**, ETH Zurich, Switzerland
Co-authors: Stefano Marelli, Paul Embrechts, Bruno Sudret

A regression method is presented for data driven problems based on polynomial chaos expansion (PCE). PCE is an established uncertainty quantification method, typically used to replace a computationally expensive (e.g., a finite element) model subject to random inputs with an inexpensive-to-evaluate multivariate polynomial. The metamodel enables a reliable estimation of the response statistics, provided that a suitable probabilistic model of the input is used. In classical machine learning (ML) regression settings, instead, the system is only known through observations of its inputs and output, and the interest lies in obtaining accurate point predictions of the latter. Here, we show that a PCE metamodel purely trained on input and output data, with input dependencies modelled through copulas, can yield point predictions whose accuracy is comparable to that of other ML methods, such as neural networks. Additionally though, the methodology enables to quantify the output uncertainty accurately, and is robust to noise. Furthermore, it enjoys additional desirable properties, such as high performance also for small training sets, and simplicity of construction, with only little parameter tuning required. In the presence of coupled inputs, we investigate two alternative ways to build the PCE, and show through simulations the superiority of one approach in the stated settings.

C0286: **Geometry of discrete copulas**

Presenter: **Elisa Perrone**, Massachusetts Institute of Technology, United States

Co-authors: Liam Solus, Caroline Uhler

Discrete copulas serve as a useful tool for modeling dependence among random variables. The space of discrete copulas admits a representation as a convex polytope which has been exploited in entropy-copula methods relevant to environmental sciences. We further analyze geometric features of discrete copulas with prescribed stochastic properties. In particular, we show that bivariate discrete copulas with a property known as ultramodularity have polytopal representations, thereby opening the door to applying linear optimization techniques in the identification of ultramodular copulas. We first draw connections to the Birkhoff polytope, alternating sign matrix polytope, and their more extensive generalizations in discrete geometry. Then, we identify the minimal collection of bounding affine inequalities of the polytope of ultramodular discrete copulas, and present techniques to construct subsets of its vertices. Finally, we discuss how to possibly exploit the introduced polytopal representations to develop new theory in applied fields, such as meteorology and climatology.

CG003 Room Orion Hall ADVANCES IN COMPOSITIONAL DATA ANALYSIS	**Chair: Alessandra Menafoglio**

C0154: **CAMPLET: Seasonal adjustment without revisions**

Presenter: **Barend Abeln**, Investment consultant, Netherlands

Co-authors: Jan Jacobs, Pim Ouwehand

The aim is to present a seasonal adjustment program called CAMPLET, an acronym of its tuning parameters, which consists of a simple adaptive procedure to separate the seasonal and the nonseasonal component from an observed time series. Once this process is carried out, there will be no need to revise these components at a later stage when new observations become available. Recently, two most widely used seasonal adjustment methods, Census X12ARIMA and TRAMO SEATS, merged into X13ARIMA SEATS to become a new industry standard. We compare and contrast CAMPLET with X13ARIMA SEATS. The main features of CAMPLET are described, and a brief review of X13ARIMASEATS is provided. We evaluate the outcomes of both methods in a controlled simulation framework using a variety of processes. We apply CAMPLET and X13ARIMA SEATS to three time series: U.S.non farm payroll employment, operational income of Ahold and real GDP in the Netherlands.

C0241: **Dealing with count zeros in compositional data analysis using the logratio-normal-multinomial distribution**

Presenter: **Marc Comas-Cufi**, Universitat de Girona, Spain

Co-authors: Josep Antoni Martin-Fernandez, Gloria Mateu-Figueras, Javier Palarea-Albaladejo

Multivariate count data are commonly modelled using the multinomial distribution. The Dirichlet distribution has been proposed for the multinomial probability parameter to account for data overdispersion. In this scenario, the resulting compound distribution is the so-called Dirichlet-multinomial (DM) distribution. Although it satisfies appealing mathematical properties, the DM distribution assumes a fairly rigid covariance structure in practice. Alternatively, the logratio-normal-multinomial (LNM) distribution is the compound probability distribution resulting from considering a multivariate logistic-normal as the distribution for the probability parameter of the multinomial distribution. Compositional analysis of multivariate count data focuses on the logratios between multinomial components and, hence, the presence of zero counts is a practical problem. We introduce a new computational treatment of count zeros based on the LNM model. A quasi-Monte Carlo EM algorithm is implemented to estimate the model parameters. The performance of our proposal will be illustrated using real and simulated data sets and compared with existing approaches.

C0283: **On analyzing zero patterns in compositional data sets**

Presenter: **Jose Antonio Martin-Fernandez**, Universitat de Girona, Spain

Co-authors: Javier Palarea-Albaladejo

Compositional Data (CoDa) are samples of random vectors representing parts of a whole, which only carry relative information. CoDa consist of vectors with strictly positive components whose sum is usually constant (e.g., 1, 100%, 106). In some applications, CoDa sets include so-called essential zeros. That is, zeros corresponding with parts genuinely absent from the composition and not with some form of censoring. For example, this is usual in time use research where some individuals spend no time on a certain activity category. Essential zeros are troublesome because it is not generally realistic to replace them by small values. To investigate whether the patterns of zeros are associated to subpopulations in the data, the subgroups of samples defined by the pattern of zeros can be analyzed in terms of compositional location and variability measures obtained from common non-zero parts. Graphical and statistical tools are introduced in this work to explore and testing for differences between groups defined by zero patterns. In particular, parametric and permutation tests for log-ratio variances are presented. These tests are further generalized for the case of projections along log-contrasts of interest determined by the user. Their performance is illustrated through real and simulated data sets.

Friday 31.08.2018 14:30 - 16:30 Parallel Session O – COMPSTAT2018

CO094 Room C3 TUTORIAL 2 **Chair: Manuel Escabias**

C0451: **An online application for functional data analysis based on R**

Presenter: **Manuel Escabias**, University of Granada, Spain

The aim is to present Statfda, an online application which allows the employment of some functional data methods based on the R 'fda package'. In spite of using functions of R, the application is made so that functional data analysis can be applied without knowing R programming. The index of the presentation is as follows: 1-Introduction to functional data analysis; 2-Information management: different possibilities of data files; 3-Basis expansion of sample curves; 3-Functional Data Analysis: Exploratory analysis, Functional PCA, Functional principal component linear regression; 4- Functional principal component logistic regression; 5-Results of the analysis: Display by screen (graphics) and download of results (text).

Saturday 01.09.2018 09:40 - 10:30 Room: C3 Chair: Frederic Ferraty CRoNoS FDA keynote talk 1

Intrinsic Riemannian functional data analysis

Speaker: **Fang Yao, Peking University, University of Toronto, China**

A new foundational framework is developed for analyzing Riemannian functional data, including intrinsic Riemannian functional principal component analysis (iRFPCA) and intrinsic Riemannian functional linear regression (iRFLR). The key concept in our development is a novel tensor Hilbert space along a curve on the manifold, based on which Karhunen-Loeve expansion for a Riemannian random process is established for the first time. This framework also features a proper comparison of objects from different tensor Hilbert spaces, which paves the way for asymptotic analysis in Riemannian functional data analysis. Built upon intrinsic geometric concepts such as vector field, Levi-Civita connection and parallel transport on Riemannian manifolds, the proposed framework embraces full generality of applications and proper handle of intrinsic geometric concepts. We then provide estimation procedures for iRFPCA and iRFLR that are distinct from their traditional and/or extrinsic counterparts, and investigate their asymptotic properties within the intrinsic geometry. Numerical performance is illustrated by simulated and real examples.

Sunday 02.09.2018 09:40 - 10:30 Room: C3 Chair: Ana Colubi CRoNoS FDA keynote talk 2

Variable selection in functional additive regression models

Speaker: **Manuel Febrero-Bande, University of Santiago de Compostela, Spain**

The problem of variable selection is considered in the case of functional variables that may be mixed with other type of variables (scalar, multivariate, directional, etc.). Our proposal begins with a simple null model and sequentially selects a new variable to be incorporated into the model based on the use of distance correlation. For the sake of simplicity, only additive models are considered. However, the proposed algorithm may assess the type of contribution (linear, non linear, ...) of each variable. The algorithm has shown quite promising results when applied to simulations and real data sets.

Friday 31.08.2018 09:00 - 10:30 Parallel Session A – CRONOSFDA2018

CI052 Room Vega Hall PRE-SUMMER SCHOOL SESSION I (OPTIONAL) Chair: Frederic Ferraty

C0199: **Functional data and nonparametric modelling: Theoretical/methodological/practical aspects**

Presenter: **Frederic Ferraty**, Mathematics Institute of Toulouse, France

Situations when one observes a response (scalar or functional variable) and functional predictor(s) are considered. The natural statistical question is very simple: are we able to predict correctly the response from the functional predictor(s) when one has no idea on the relationship between the response and functional predictor(s)? A suitable answer to this important statistical issue is the "functional nonparametric regression". The word "nonparametric" stands for any model requiring very few assumptions with respect to the relationship between the response and the predictor(s); the word "functional" reminds that the model has to handle functional data. So, the aim is to give an extensive overview on this statistical topic. In addition to some theoretical and practical key developments, real dataset illustrate the purpose (benchmark datasets, hyperspectral image, forensic entomology in the context of criminology, etc).

Friday 31.08.2018 11:00 - 12:00 Parallel Session B – CRONOSFDA2018

CI054 Room Cuza Hall PRE-SUMMER SCHOOL SESSION II (OPTIONAL) **Chair: Fang Yao**

C0200: **Special Invited Session: Advances in functional data analysis**

Presenter: **Fang Yao**, Peking University, University of Toronto, China

Co-authors: Aymeric Stamm, Michelle Carey

The special invited session has been organized by Laura Sangalli for COMPSTAT 2018 and it consists of 3 research talks: 1- Inference in separable Hilbert spaces using Hotelling's T2 (Presented by A. Stamm); 2- Intrinsic Riemmanian functional data analysis (Presented by F. Yao); 3- Dynamic Modelling with Data2PDE (Presented by M. Carey).

Friday 31.08.2018 14:30 - 16:30 Parallel Session C – CRONOSFDA2018

CI028 Room C3 SUMMER SCHOOL SESSION I — Chair: Manuel Escabias

C0187: **An online application for functional data analysis based on R**

Presenter: **Manuel Escabias**, University of Granada, Spain

The aim is to present Statfda, an online application which allows the employment of some functional data methods based on the R 'fda package'. In spite of using functions of R, the application is made so that functional data analysis can be applied without knowing R programming. The index of the presentation is as follows: 1-Introduction to functional data analysis; 2-Information management: different possibilities of data files; 3-Basis expansion of sample curves; 3-Functional Data Analysis: Exploratory analysis, Functional PCA, Functional principal component linear regression; 4- Functional principal component logistic regression; 5-Results of the analysis: Display on screen (graphics) and results downloading (text).

CO026 Room C413 FUNCTIONAL SPATIO-TEMPORAL DATA AND APPLICATIONS — Chair: Sophie Dabo

C0163: **On the maximal norm of the functional periodogram**

Presenter: **Clement Cerovecki**, Univ libre de Bruxelles, Belgium

Co-authors: Siegfried Hormann, Vaidotas Characiejus

The periodogram of multivariate and functional data is considered and the limiting distribution of the maximal norm over fundamental frequencies is derived. We provide conditions which assure that this maximum is in the domain of attraction of the Gumbel distribution. To this end, we use a Gaussian approximation. Our results generalize a previous theorem to multivariate and functional data. We propose an application to test for hidden periodic patterns in functional time series. In particular, we extend a recent test.

C0179: **Modelization of the tropical atlantic sea surface temperatures using spatial functional principal component analysis**

Presenter: **Ogoudjobi Francois Adjibode**, CIPMA, Benin

A spatial functional principal component analysis is presented, and applications to two types (monthly and daily) of spatio-temporal sea surface temperature (SST)data from a given basin are given. We compare our results to those obtained with the linear inversion model used for empirical orthogonal functions (EOF) analysis. The proposed functional spatio-time series method allows us to put out the most dominant mode of the basin, and to propose a suitable model for understanding the interaction between the air and the ocean.

C0180: **Clustering spatial functional data**

Presenter: **Vincent Vandewalle**, Inria, France

Co-authors: Cristian Preda, Sophie Dabo

Two approaches for clustering spatial functional data are presented. The first one is the model-based clustering that uses the concept of density for functional random variables and logistic weights on the prior cluster probabilities depending on spatial coordinates. The second one is the hierarchical clustering based on univariate statistics for functional data such as the functional mode or the functional mean, and includes spatial weights in the distances computation. These two approaches take into account the spatial features of the functional data: two observations that are spatially close share a common distribution of the associated random variables. The two methodologies are illustrated by an application to air quality data.

C0178: **Forecasting multiple functional time series: A static factor approach**

Presenter: **Gilles Nisol**, ULB, Belgium

Co-authors: Siegfried Hormann, Marc Hallin

Theoretical foundations and a practical method to forecast multiple functional time series (FTS) are set. In order to do so, we generalize the static factor model to the case where cross-section units are FTS. We first derive a representation result. We show that if the K first eigenvalues of the covariance operator of the cross-section of the N FTS are unbounded while N grows and if the $K+1$ eigenvalue is bounded, then we can represent the FTS as a sum of a common component driven by K factors and an idiosyncratic component. We then set up an information criterion that chooses jointly the number K of factors and the dimension on which we should project the FTS before estimating the static factor model. We suggest a method of estimation and prediction based on these projected FTS. We assess the performances of the method and information criterion through a simulation exercise. Finally, we consider a real-data application. We show that by applying our method to a cross-section of PM10 concentration curves obtained across several measurement centers in Graz, we have a better prediction accuracy than by limiting the analysis to individual FTS.

C0159: **White noise testing for functional time series: Application to model identification and diagnosis**

Presenter: **Guillermo Mestre**, Universidad Pontificia Comillas, Spain

Co-authors: Jose Portela, Antonio Munoz, Estrella Alonso

White noise characterization is a crucial step in the identification and diagnosis of a model for scalar time series, where the autocorrelation and partial autocorrelation functions of the time series are the most common tools used for this purpose. An autocorrelation function for functional time series is proposed, based on the L^2 norm of the lagged covariance operators of the time series. The distribution of this sequence of statistics has been established under the assumption of functional white noise, hence providing a method to test the adequacy of functional time series models by checking if the residuals of a fitted model do not exhibit serial autocorrelation. This method is validated by numerical simulations of both white noise and dependent functional processes, where the structure of the process is identified by its autocovariance norms and a linear model is fitted and diagnosed using the described techniques. The applicability of the method is illustrated via an application to two real-world datasets, including Spanish electricity prices profiles.

Friday 31.08.2018 17:00 - 18:00 Parallel Session D – CRONOSFDA2018

CI030 Room C413 SUMMER SCHOOL SESSION II Chair: Manuel Febrero-Bande

C0188: **Computational aspects on functional data analysis**

Presenter: **Manuel Febrero-Bande**, University of Santiago de Compostela, Spain

This course reviews some of the main techniques in Functional Data Analysis focusing on its computational aspects. The course covers the following list of the topics: Representation, Simulation, Exploratory Data Analysis, Regression, Classification and Testing providing examples of use using the R-package fda.usc.

CO018 Room C3 ROBUST FUNCTIONAL DATA ANALYSIS Chair: Stefan Van Aelst

C0174: **Robust functional principal components with sparse observations**

Presenter: **Matias Salibian-Barrera**, The University of British Columbia, Canada

Co-authors: Graciela Boente, Jane-Ling Wang

Principal components analysis provides an optimal lower-dimensional approximation to multivariate observations. Similarly, functional principal components analysis may yield parsimonious predictions for each trajectory in the sample. A new characterization of elliptical distributions on separable Hilbert spaces shows that this holds even when second moments do not exist. We discuss the problem of robust estimation of functional principal components when only a few observations are available per curve. The conditional expectation approach estimates the covariance function by smoothing the sparsely available cross-products, and thus "combines information" from many sparse curves. A first attempt at protecting this approach from outliers by using a robust smoother does not work because the distribution of the cross-products is generally asymmetric. However, when the stochastic process has an elliptical distribution, one can exploit the linear structure of the conditional distribution of the process at time t conditional on its value at time s to obtain robust estimators of the scatter function $G(t,s)$.

C0162: **Robust functional regression based on principal components**

Presenter: **Ioannis Kalogridis**, KU Leuven, Belgium

Functional data analysis is a fast evolving branch of modern statistics, yet despite the popularity of the functional linear model in recent years, almost all estimation methods rely on generalized least squares procedures and as such are sensitive to atypical observations. To remedy this, we propose a two-step estimation procedure that combines robust functional principal components and robust linear regression. We further propose a transformation that reduces the curvature of the estimates and can be advantageous in many settings. For these methods we prove Fisher-consistency for elliptical distributions and consistency under mild regularity conditions. Simulation experiments show that the proposed estimators have reasonable efficiency, protect against outlying observations, produce smooth estimates and compare favourably with the few existing robust approaches.

C0185: **Robust change point procedures for functional data**

Presenter: **Alexander Duerre**, TU Dortmund, Germany

If functional data is observed over time, one is often confronted with the question whether its underlying distribution changes at one or several time points. The usual change point procedures have a linear structure. It is noticed for one-dimensional time series that such tests are heavily influenced by outliers which can either mask a structural change or pretend a change point. Furthermore they behave poorly under heavy tailed distributions. Therefore, robust alternatives for change point detection of functional data are presented. We apply these methods to brain activity data and respiratory air data.

Friday 31.08.2018 18:00 - 19:00 Parallel Session E – CRONOSFDA2018

CI032 Room C413 SUMMER SCHOOL SESSION III **Chair: Manuel Febrero-Bande**

C0189: **Computational aspects on functional data analysis**

Presenter: **Manuel Febrero-Bande**, University of Santiago de Compostela, Spain

This course reviews some of the main techniques in Functional Data Analysis focusing on its computational aspects. The course covers the following list of the topics: Representation, Simulation, Exploratory Data Analysis, Regression, Classification and Testing providing examples of use using the R-package fda.usc.

Saturday 01.09.2018 08:30 - 09:30 Parallel Session F – CRONOSFDA2018

CI034 Room C413 SUMMER SCHOOL SESSION IV Chair: Manuel Febrero-Bande

C0190: **Computational aspects on functional data analysis**

Presenter: **Manuel Febrero-Bande**, University of Santiago de Compostela, Spain

This course reviews some of the main techniques in Functional Data Analysis focusing on its computational aspects. The course covers the following list of the topics: Representation, Simulation, Exploratory Data Analysis, Regression, Classification and Testing providing examples of use using the R-package fda.usc.

CO004 Room C3 PATCHWORK OF FDA Chair: Frederic Ferraty

C0152: **Exploration of shape features for functional data**

Presenter: **Stanislav Nagy**, Charles University, Czech Republic

In many situations, the shape of functional observations is an important feature that must be taken into account in statistical analysis. The information about the shape properties can be extracted from the derivatives of the sample trajectories. Though, this approach can be applied only if the curves are regular and smooth, and the derivatives must be estimated. We present a simple alternative to this methodology based on simultaneous evaluation of multivariate projections of the data. This technique does not require smoothness or continuity, yet provides fine recognition of shape traits of the curves. The idea is illustrated on - but not limited to - functional data depth.

C0154: **A robust t-process regression model with independent errors**

Presenter: **Jian Qing Shi**, Newcastle University, United Kingdom

Gaussian process regression (GPR) model is well-known to be susceptible to outliers. Robust process regression models based on t-process or other heavy-tailed processes have been developed to address the problem. However, due to the nature of the current definition for heavy-tailed processes, the unknown process regression function and the random errors are always defined jointly and thus dependently. This definition, mainly owing to the dependence assumption involved, is not justified in many practical problems and thus limits the application of those robust approaches. It also results in a limitation of the theory of robust analysis. We will discuss a new robust process regression model enabling independent random errors and will also discuss an efficient estimation procedure. We will present an application to analyse a medical game data and show that the proposed method is robust against outliers and has a better performance in prediction compared with the existing models.

C0160: **Points of impact in generalized linear models with functional predictors**

Presenter: **Dominik Liebl**, University Bonn, Germany

Co-authors: Dominik Poss

Generalized linear models with function-valued predictor variables and scalar outcomes belong to the well-established and widely used methodological toolbox in functional data analysis. In many applications, however, only specific locations or time-points of the functional predictors have an impact on the outcome. The selection of such points of impact constitutes a particular variable selection problem, since the high correlation in the functional predictors violates the basic assumptions of existing high-dimensional variable selection procedures. We introduce a generalized linear regression model with functional predictors evaluated at unknown points of impact which need to be estimated from the data alongside the model parameters. We propose a threshold-based and a fully data-driven estimator, establish the identifiability of our model, derive the convergence rates of our point of impact estimators, and develop the asymptotic normality of the estimators of the linear model parameters. The finite sample properties of our estimators are assessed by means of a simulation study. Our methodology is motivated by a psychological case study in which the participants were asked to continuously rate their emotional state while watching an affective online video on the persecution of African albinos.

Saturday 01.09.2018 11:00 - 13:00 Parallel Session H – CRONOSFDA2018

CI036 Room C413 SUMMER SCHOOL SESSION V Chair: Manuel Febrero-Bande

C0191: **Computational aspects on functional data analysis**
Presenter: **Manuel Febrero-Bande**, University of Santiago de Compostela, Spain
This course reviews some of the main techniques in Functional Data Analysis focusing on its computational aspects. The course covers the following list of the topics: Representation, Simulation, Exploratory Data Analysis, Regression, Classification and Testing providing examples of use using the R-package fda.usc.

CO008 Room C3 RECENT ADVANCES ON FUNCTIONAL DATA ANALYSIS AND APPLICATIONS Chair: Ana Maria Aguilera

C0157: **SFLM: A mix of a Functional Linear Model and of a Spatial autoregressive model for spatially correlated functional data**
Presenter: **Gilbert Saporta**, CNAM, France
Co-authors: Tingting Huang, Huiwen Wang, Shanshan Wang
The well-known functional linear regression model (FLM) has been developed under the assumption that the observations are independent. However, the independence assumption may often be violated in practice, especially when we collect data with network structure coming from various fields such as marketing, sociology or spatial economics. Yet relatively few works are available for FLM with network structure. We propose a novel spatial functional linear model (SFLM), incorporating a spatial autoregressive parameter and a spatial weight matrix in FLM to accommodate spatial dependence among individuals. The proposed model is flexible as it takes advantages of FLM in dealing with high dimensional covariates, and of spatial autoregressive model (SAR model) in capturing network dependence. We develop an estimation method based on functional principal components analysis (FPCA) and maximum likelihood estimation. The simulation studies show that our method performs as well as FPCA-based method for FLM when there is no network structure and outperforms the latter when there exists a network structure. A real dataset of weather data is also employed to demonstrate the utility of SFLM.

C0168: **Functional tools for increasing the accuracy of biodiversity assessment**
Presenter: **Tonio Di Battista**, G. d'Annunzio University of Chieti-Pescara, Italy, Italy
Co-authors: Francesca Fortuna, Fabrizio Maturo
Biodiversity is recognized as one of the most important indicators for environmental assessment. However, no scientific consensus has been reached about how to properly measure and monitor it. This is mainly due to the multivariate nature of biodiversity. To overcome this issue, we propose a new methodological approach for monitoring biodiversity introducing a functional approach to diversity profiles. Indeed, the latter may be naturally considered as functional data because they are expressed as functions of the species abundance vector in a fixed domain. Specifically, several functional tools are developed such as the derivatives, the radius of curvature, the curve length, the biodiversity surface, and the volume under the surface. Each functional tool reflects a specific aspect of biodiversity. Thus, the combined use of them provides a useful method for identifying areas of high environmental risk, with the potential to address the monitoring of environmental policies. The main purpose of this research is to provide specialists and scholars with additional tools to improve the understanding of the dynamics of biodiversity.

C0170: **The control of family-wise error rate for functional data: A unified framework**
Presenter: **Alessia Pini**, Universita Cattolica del Sacro Cuore, Italy
Co-authors: Konrad Abramowicz, Lina Schelin, Sara Sjostedt de Luna, Aymeric Stamm, Simone Vantini
Inference for functional data is currently approached in two different ways: global inference aiming at testing functional hypotheses over the entire domain, and local inference aiming at selecting domain subsets responsible for the rejection of a null hypothesis. In the local setting, a p-value can be computed at every point of the domain, obtaining an unadjusted p-value function, which controls only pointwise the probability of type I error: for all points, the probability of type I error is controlled, but the probability of committing at least one type I error (i.e., the so-called familywise error rate - FWER) is not. Hence, the unadjusted p-value function cannot be used for domain selection purposes, and adjusted p-value functions are needed. A unified framework for methods that fit this purpose is presented. It includes and extends existing methods and our own proposed one. Their inferential properties are characterized in terms of finite-sample or asymptotic control of the FWER and consistency. Finite-sample properties are compared on a simulation study. Finally, the proposed local inferential techniques are applied to knee kinematic and brain tractography data.

C0173: **Strongly-consistent prediction of air pollutants PM10 by high-singular ARBX(1) processes**
Presenter: **Javier Alvarez-Liebana**, University of Granada, Spain
Co-authors: Maria Dolores Ruiz-Medina
Our main motivation relies on prediction of daily mean concentrations curves of air pollutants PM10 (coarse particles), in the Haute-Normandie region (France), including exogenous meteorological variables. Due to the lack of proposals on the statistical analysis, and prediction, based on functional linear time series theory, with exogeneous singular functional random variables, we provide a methodological framework to address this problem, based on wavelet bases and Besov spaces of negative order. A simulation study is undertaken, to illustrate the asymptotic properties of the ARBX(1) componentwise estimator formulated for the functional correlation structure, and of its associated plug-in predictor. A real-data application is developed for functional prediction of PM10 air pollutants.

C0181: **A novel approach to domain selection in functional data: Boosting classification performance**
Presenter: **Nicolas Hernandez**, Universidad Carlos III de Madrid, Spain
Co-authors: Alberto Munoz, Gabriel Martos
A domain selection approach is proposed for classification problems in functional data. Consider two samples of random elements $f_1,\ldots,f_n$ and $g_1,\ldots,g_m$ in $L^2(X)$ generated from the functional stochastic models $f_i(x)=\mu_k(x)+\varepsilon_i(x)$ for $i=1,\ldots,n$ and $g_j(x)=\mu_k(x)+\varepsilon_j(x)$ for $j=1\ldots,m$ respectively and defined on the same domain $X=[0,1]$. The function $\mu_k(x)$ is the mean function for $k=f,g$ and $\varepsilon(x)$ is a random and independent functional error that captures the variability within each class. In this setting, we propose to use a local–inner product parametrized by the vector $\theta=(\theta_1,\theta_2)$, with $0\leqslant\theta_1<\theta_2\leqslant 1$, such that, $\langle f,g\rangle_\theta=\int_{\theta_1}^{\theta_2} f(x)g(x)dx$. The proposed inner–product induce a local–metric in the space of random elements $L^2(X)$. The optimization of θ is presented as a domain selection technique, where the optimization goal pursue the minimization of the misclassification error rate when classifying samples of random functions.

Saturday 01.09.2018 14:30 - 16:00 Parallel Session I – CRONOSFDA2018

CI038 Room C413 SUMMER SCHOOL SESSION VI Chair: Manuel Febrero-Bande

C0192: **Computational aspects on functional data analysis**

Presenter: **Manuel Febrero-Bande**, University of Santiago de Compostela, Spain

This course reviews some of the main techniques in Functional Data Analysis focusing on its computational aspects. The course covers the following list of the topics: Representation, Simulation, Exploratory Data Analysis, Regression, Classification and Testing providing examples of use using the R-package fda.usc.

CO012 Room C3 FUNCTIONAL DATA ANALYSIS: THEORY AND APPLICATIONS Chair: Enea Bongiorno

C0158: **Kriging spatial functional data over complex domains through random domain decompositions**

Presenter: **Alessandra Menafoglio**, Politecnico di Milano, Italy

Co-authors: Piercesare Secchi, Giorgia Gaetani

The analysis of complex data distributed over large or highly textured regions poses new challenges for spatial statistics. We present a novel methodology for the spatial prediction of object data distributed over such complex regions, which deals with the data and the domain complexities through a divide-et-impera approach. We propose to perform repeated Random Domain Decompositions, each defining a set of homogeneous sub-regions where to perform local object-oriented spatial analyses, under stationarity assumptions, to be then aggregated into a final global analysis. The method we propose is entirely general, and prone to be used with numerous types of object data (e.g., functional data, density data or manifold data), being grounded upon the theory of Object Oriented Spatial Statistics. As an insightful illustration of the method, we consider the spatial prediction of the PDF of dissolved oxygen in the estuarine systems of the Chesapeake Bay (US).

C0175: **Procrustes metrics on covariance operators and optimal transportation of Gaussian processes**

Presenter: **Yoav Zemel**, Georg-August Universitat Gottingen, Germany

Co-authors: Victor Panaretos, Valentina Masarotto

Covariance operators are fundamental in functional data analysis, providing the canonical means to analyse functional variation via the celebrated Karhunen–Loeve expansion. These operators may themselves be subject to variation, for instance in contexts where multiple functional populations are to be compared. Statistical techniques to analyse such variation are intimately linked with the choice of metric on covariance operators, and the intrinsic infinite-dimensionality of these operators. We describe the manifold-like geometry of the space of trace-class infinite-dimensional covariance operators and associated key statistical properties, under the recently proposed infinite-dimensional version of the Procrustes metric. We identify this space with that of centred Gaussian processes equipped with the Wasserstein metric of optimal transportation. The identification allows us to provide a detailed description of those aspects of this manifold-like geometry that are important in terms of statistical inference; to establish key properties of the Frechet mean of a random sample of covariances; and to define generative models that are canonical for such metrics and link with the problem of registration of warped functional data.

C0177: **Functional insights into Google AdWords**

Presenter: **Christoph Rust**, University of Regensburg, Germany

Co-authors: Dominik Liebl, Stefan Rameseder

The functional linear regression model with points of impact is a recent augmentation of the classical functional linear model with many practically important applications. However, we demonstrate that the existing procedure for estimating the parameters of this regression model can be very inaccurate for practical sample sizes. A particularly problematic aspect is the tendency to omit relevant points of impact resulting in omitted variable biases. Therefore, we provide an adjusted estimation algorithm which is designed to compensate for this impractical small sample behavior. Our estimation algorithm is compared with the existing estimation procedure using extensive Monte Carlo simulations. Applicability of our estimation algorithm is demonstrated using data from Google AdWords, todays most important platform for online advertisements.

C0183: **On the average excess for expectations of L2-valued random elements**

Presenter: **Gil Gonzalez-Rodriguez**, University of Oviedo, Spain

Co-authors: Ana Belen Ramos-Guajardo

Several procedures for checking the equality of means of random elements on a separable Hilbert space are available. Once the difference of expectations can be assessed at a given significance level, additional analysis are needed in order to explain such a difference. When dealing with real random variables, one-side hypothesis tests are usually considered to this aim. Nevertheless, when the space lacks of total ordering, such an approach is usually not applicable. This is the typical situation in the functional context, where frequently an L2 space finite-measurable with respect to the Lebesgue measure is considered as a framework. In this situation, when two random functions have different expectations, part of one expectation may excess the other and vice-versa. In such a case, the average excess of one expectation over the other may be useful. Given two random functions X and Y taking on values in a finite-measure L2 space, the aim is to develop inferences about the average excess of $E(X)$ over $E(Y)$. A centred empirical version of the average excess will be considered and its asymptotic distribution will be derived under mild conditions. Applications in different contexts will be shown.

Saturday 01.09.2018 16:30 - 18:00 Parallel Session J – CRONOSFDA2018

CI040 Room C413 SUMMER SCHOOL SESSION VII Chair: Manuel Febrero-Bande

C0193: **Computational aspects on functional data analysis**

Presenter: **Manuel Febrero-Bande**, University of Santiago de Compostela, Spain

This course reviews some of the main techniques in Functional Data Analysis focusing on its computational aspects. The course covers the following list of the topics: Representation, Simulation, Exploratory Data Analysis, Regression, Classification and Testing providing examples of use using the R-package fda.usc.

CO006 Room C3 NONPARAMETRIC ANALYSIS OF FUNCTIONAL DATA Chair: Stanislav Nagy

C0151: **Estimation of temperature-dependent growth profiles of fly larvae with application to criminology**

Presenter: **Frederic Ferraty**, Mathematics Institute of Toulouse, France

It is not unusual in cases where a body is discovered that it is necessary to determine a time of death or more formally a post mortem interval (PMI). Forensic entomology can be used to estimate this PMI by examining evidence obtained from the body from insect larvae growth. We propose a method to estimate the hatching time of larvae (or maggots) based on their lengths, the temperature profile at the crime scene and experimental data on larval development. This requires the estimation of a time-dependent growth curve from experiments where larvae have been exposed to a relatively small number of constant temperature profiles. Since the temperature influences the developmental speed, a crucial step is the time alignment of the curves at different temperatures. We then propose a model for time varying temperature profiles based on the local growth rate estimated from the experimental data. This allows us to find out the most likely hatching time for a sample of larvae from the crime scene. We explore via simulations the robustness of the method to errors in the estimated temperature profile and apply it to the data from two criminal cases from the United Kingdom. Asymptotic properties are also provided for the estimators of the growth curves and the hatching time.

C0153: **On nonparametric depth based classification of functional observations**

Presenter: **Pauliina Ilmonen**, Aalto University School of Science, Finland

Co-authors: Sami Helander, Stanislav Nagy, Germain Van Bever, Lauri Viitasaari

The aim is to discuss assessing typicality of functional observations. Moreover, we provide a new classification method that is based on j-th order k-th moment integrated depths. For $j = 1$ and $k = 1$ this is equal to applying the mean halfspace depth of a functional value with respect to the corresponding univariate marginal distribution. When j is larger than 1, the method is not based on comparing location only but considers shape of the function as well. Moreover, the method can be applied to partially observed functions without extrapolation or interpolation. Theoretical properties of the new approach are explored and several real data examples are presented to demonstrate its excellent classification performance.

C0155: **On Mahalanobis distance in functional settings**

Presenter: **Beatriz Bueno-Larraz**, Universidad Autonoma de Madrid, Spain

Co-authors: Jose Berrendero, Antonio Cuevas

The theory of Reproducing Kernel Hilbert Spaces (RKHS's) has found many interesting applications in different fields, including statistic. For instance, it helps to partially overcome some difficulties that arise when moving from the multivariate context to the functional one, like the non-invertivility of the covariance operators. One of the problems derived from this non-invertivility is that it does not exist a functional counterpart of the Mahalanobis distance (a relevant notion of multivariate depth). We suggest a suitable functional version of this distance based on the RKHS associated with the underlying stochastic process of the data. This new statistical distance inherits some interesting properties of the original multivariate distance and has shown good performances in different problems (like functional classification, outlier detection, etc).

C0171: **Regularized classifiers of functional data under partial observation**

Presenter: **David Kraus**, Masaryk University, Czech Republic

Co-authors: Marco Stefanucci

Classification of functional data into two groups by linear classifiers is considered on the basis of one-dimensional projections of functions. We reformulate the task to find the best classifier as an optimization problem and solve it by regularization techniques, namely the conjugate gradient method with early stopping, the principal component method and the ridge method. We study the empirical version with finite training samples consisting of incomplete functions observed on different subsets of the domain and show that the optimal, possibly zero, misclassification probability can be achieved in the limit along a possibly non-convergent empirical regularization path. Being able to work with fragmentary training data we propose a domain extension and selection procedure that finds the best domain beyond the common observation domain of all curves. In a simulation study we compare the different regularization methods and investigate the performance of domain selection. Our methodology is illustrated on a medical data set, where we observe a substantial improvement of classification accuracy due to domain extension.

Saturday 01.09.2018 18:00 - 19:00 Parallel Session K – CRONOSFDA2018

CI042 Room C413 SUMMER SCHOOL SESSION VIII **Chair: Fang Yao**

C0194: **Functional data analysis and related topics**

Presenter: **Fang Yao**, Peking University, University of Toronto, China

Functional data analysis (FDA) has received substantial attention in recent years, with applications arising from various disciplines, such as engineering, public health, finance etc. In general, the FDA approaches focus on nonparametric underlying models that assume the data are observed from realizations of stochastic processes satisfying some regularity conditions. The estimation and inference procedures usually do not depend on just a finite number of parameters, which contrasts with parametric models, and exploit techniques such as smoothing methods, dimension reduction that allow data to speak for themselves. This tutorial will cover general ideas in functional data analysis, such as functional principal component analysis, basis representation models, functional linear regression as well as more flexible regression type models, and so on. Some basic computing and data analysis using R and/or Matlab will be also introduced.

Sunday 02.09.2018 08:30 - 09:30 Parallel Session L – CRONOSFDA2018

CI044 Room C413 SUMMER SCHOOL SESSION IX Chair: Fang Yao

C0195: **Functional data analysis and related topics**

Presenter: **Fang Yao**, Peking University, University of Toronto, China

Functional data analysis (FDA) has received substantial attention in recent years, with applications arising from various disciplines, such as engineering, public health, finance etc. In general, the FDA approaches focus on nonparametric underlying models that assume the data are observed from realizations of stochastic processes satisfying some regularity conditions. The estimation and inference procedures usually do not depend on just a finite number of parameters, which contrasts with parametric models, and exploit techniques such as smoothing methods, dimension reduction that allow data to speak for themselves. This tutorial will cover general ideas in functional data analysis, such as functional principal component analysis, basis representation models, functional linear regression as well as more flexible regression type models, and so on. Some basic computing and data analysis using R and/or Matlab will be also introduced.

CO022 Room C3 FUNCTIONAL DATA WITH SPATIAL DEPENDENCE Chair: Alessandra Menafoglio

C0156: **Spatial smoothing of Raman spectra sampled on a regular grid**

Presenter: **Kevin Hayes**, University of Limerick, Ireland

Co-authors: Darren Whitaker

The pharmaceutical manufacturing industry routinely uses advanced soft sensor technologies to monitor and control product quality. For the data under consideration, the pharmaceutical ingredients were blended and compacted into 12 mm diameter tablets, and on each tablet 407 individual spectra in the range 1230 to 1330 cm^{-1} were recorded at 500 μm intervals on a regular lattice. The objective is to model the spatial dependence between these intrinsically functional data measurements and indirectly evaluate the spatial distribution of the active pharmaceutical ingredient through the tablet. Also of interest is the efficiency of sampling protocol used. We apply linear dynamic smoothing to the spectra individually and submit the estimated parameters of the associated (second-order) differential equations to treatment by classical multivariate geostatistical methodologies.

C0165: **Modelling spatio-temporal dependent functional data via regression with differential regularization**

Presenter: **Eleonora Arnone**, Politecnico di Milano, Italy

Co-authors: Laura Azzimonti, Laura Sangalli, Fabio Nobile

A new method is proposed for the analysis of functional data defined over spatio-temporal domains. These data can be interpreted as time evolving surfaces or spatially dependent curves. The proposed method is based on regression with differential regularization. We are in particular interested to the case when prior knowledge on the phenomenon under study is available. The prior knowledge is described in terms of a time-dependent Partial Differential Equation (PDE) that jointly models the spatial and temporal variation of the phenomenon. We consider various samplings designs, including geo-statistical and areal data. We show that the corresponding estimation problem is well posed and can be discretized in space by means of the Finite Element method, and in time by means of the Finite Difference method. The model can handle data distributed over spatial domains having complex shapes, such as domains with strong concavities and holes. Moreover, various types of boundary conditions can be considered. The proposed method is compared to existing techniques for the analysis of spatio-temporal models, including space-time kriging and methods based on thin plate splines and soap film smoothing. As a motivating example, we study the blood flow velocity field in the common carotid artery, using data from Echo-Color Doppler.

C0172: **Nonparametric clustering of dependent functional data with applications to climate reconstruction**

Presenter: **Konrad Abramowicz**, Umea University, Sweden

Co-authors: Lina Schelin, Sara Sjostedt de Luna, Johan Strandberg

An approach used to cluster time- and space-dependent functional data is presented. Assume that for a given spatial location there is a lattice of time points (e.g., years), where a function is observed in each time point. Further, assume that there are latent (unobservable) groups of functions that vary slowly over time, and where different groupings may arise at different time scales (resolutions). Groups are characterised by distinct frequencies of the observed functions. We propose and discuss a non-parametric double clustering method, which identifies latent groups at different resolutions. Additionally, we consider the aspect of dependency by simultaneously analysing time dependent curves at different spatial locations. The introduced methodology is applied to sediment data from three varved lakes from different parts of Scandinavia, aiming at reconstructing winter climatic regimes in the region.

Sunday 02.09.2018 11:00 - 12:30 Parallel Session N – CRONOSFDA2018

CI048 Room C413 SUMMER SCHOOL SESSION XI — Chair: Fang Yao

C0197: **Functional data analysis and related topics**

Presenter: **Fang Yao**, Peking University, University of Toronto, China

Functional data analysis (FDA) has received substantial attention in recent years, with applications arising from various disciplines, such as engineering, public health, finance etc. In general, the FDA approaches focus on nonparametric underlying models that assume the data are observed from realizations of stochastic processes satisfying some regularity conditions. The estimation and inference procedures usually do not depend on just a finite number of parameters, which contrasts with parametric models, and exploit techniques such as smoothing methods, dimension reduction that allow data to speak for themselves. This tutorial will cover general ideas in functional data analysis, such as functional principal component analysis, basis representation models, functional linear regression as well as more flexible regression type models, and so on. Some basic computing and data analysis using R and/or Matlab will be also introduced.

CO020 Room C3 NEW CHALLENGES IN FDA APPLICATIONS — Chair: Tonio Di Battista

C0164: **On the use of functional data analysis in asthma diagnosis**

Presenter: **Sara Fontanella**, Imperial College London, United Kingdom

Co-authors: Lesley Lowe, Clare Murray, Angela Simpson, Adnan Custovic

Asthma is among the most common chronic diseases in both children and adults. It embodies a multifactorial and heterogeneous condition with many different pathways of disease development and progression. This intrinsic heterogeneity represents a key confound to disease understanding. In recent decades, many longitudinal observational birth cohort studies have been developed with the aim of recording the natural history of asthma-related symptoms over time. They offer a unique opportunity to disentangle disease heterogeneity and to comprehend the underlying disease mechanisms, predict disease course, and, ultimately, design personalized treatment strategies. Wheezing is a recognizable asthma symptom and several studies attempted at characterizing asthma heterogeneity by investigating manifest patterns of wheeze over time. Within the context of the MAAS birth cohort study, we aim at characterizing wheeze progression and its association with lung function growth and asthma development. Functional data analysis provides useful tools for analyzing longitudinal data, however, the available data present unique challenges: some are non-Gaussian, the timings of the repeated measurements are sparse and irregular. To overcome these issues, we adopt a contemporary nonparametric functional analytic approach based on principal analysis through conditional expectation (PACE) coupled with a latent Gaussian process model.

C0166: **Monitoring the spatial correlation among functional data streams**

Presenter: **Stefano Antonio Gattone**, University G. d'Annunzio of Chieti-Pescara, Italy

Co-authors: Antonio Balzanella, Tonio Di Battista, Rosanna Verde, Elvira Romano

The focus is on measuring the spatial correlation among functional data streams recorded by sensor networks. In many real world applications, spatially located sensors are used for performing at a very high frequency, repeated measurements of some variable. Due to the spatial correlation, sensed data are more likely to be similar when measured at nearby locations rather than in distant places. In order to monitor such correlation over time and to deal with huge amount of data, we propose a strategy based on computing the well-known Moran's index on summaries of the data.

C0169: **Data stream reduction via functional time series analysis**

Presenter: **Francesca Fortuna**, G.d'Annunzio University of Chieti-Pescara, Italy

Co-authors: Fabrizio Maturo

Nowadays the analysis of big data streaming is of main interest in several fields, as it represents an important source of information, which may be useful for forecasting tasks. However, dealing with this type of data involves a series of challenges concerning software, format, and dimensionality issues. Indeed, a stream is an unbounded, ordered sequence of objects that can be read only once or a small number of times. The main characteristics of data streaming are that data continuously flow, and their size is extremely large and potentially infinite. In this context, extracting relevant and reliable information from big data become a crucial aspect. To this end, we analyze data streams in a functional framework, focusing on the forecasting problem in time series with nonparametric techniques. Specifically, we investigate data streams in user defined time periods, identifying a suitable probability distribution function able to describe the main characteristics of the data. In particular, we aim to study the repartition function of the stream random variable to obtain a memory of the process over time. In this framework, the high dimensionality of the data is reduced into a matrix of functional observations, whose units are represented by the user-defined time periods.

C0176: **About the complexity of a functional data set**

Presenter: **Enea Bongiorno**, Universita del Piemonte Orientale, Italy

Co-authors: Aldo Goia, Philippe Vieu

Consider the problem to state the compatibility of observed functional data with a reference model. Starting from the small ball probability factorization, it is possible to introduce the concept of complexity for functional data and suitable indexes measuring it. At a first stage, a descriptive approach, mainly based on a new graphical tool (namely the log-Volugram), is implemented and fruitfully applied. From an inferential perspective, a hypothesis test is implemented: the test statistic is derived, its asymptotic law is studied, a study of level and power of the test for finite sample sizes and a comparison with a competitor are carried out by Monte Carlo simulations. It turns out that the developed methodologies are fully free from assumptions on model, distribution as well as dominating measure. Applications are provided over financial time series.

Sunday 02.09.2018 14:00 - 15:30 Parallel Session O – CRONOSFDA2018

CI046 Room C413 SUMMER SCHOOL SESSION X **Chair: Fang Yao**

C0196: **Functional data analysis and related topics**

Presenter: **Fang Yao**, Peking University, University of Toronto, China

Functional data analysis (FDA) has received substantial attention in recent years, with applications arising from various disciplines, such as engineering, public health, finance etc. In general, the FDA approaches focus on nonparametric underlying models that assume the data are observed from realizations of stochastic processes satisfying some regularity conditions. The estimation and inference procedures usually do not depend on just a finite number of parameters, which contrasts with parametric models, and exploit techniques such as smoothing methods, dimension reduction that allow data to speak for themselves. This tutorial will cover general ideas in functional data analysis, such as functional principal component analysis, basis representation models, functional linear regression as well as more flexible regression type models, and so on. Some basic computing and data analysis using R and/or Matlab will be also introduced.

CO024 Room C3 CLUSTERING AND CLASSIFICATION FOR FUNCTIONAL DATA **Chair: Gil Gonzalez-Rodriguez**

C0161: **PCA-based discrimination of partially observed functional data, with an application to Aneurisk65 dataset**

Presenter: **Marco Stefanucci**, University of Rome - Sapienza, Italy

Co-authors: Laura Sangalli, Pierpaolo Brutti

Functional data are usually assumed to be observed on a common domain. However, it is often the case that some portion of the functional data is missing for some statistical units, invalidating most of the existing techniques for functional data analysis. The developments of methods able to handle partially observed or incomplete functional data is currently attracting an increasing interest. We briefly review this literature. We then focus on discrimination based on principal component analysis, and illustrate a few possible methods via simulation studies and an application to the AneuRisk65 dataset. We show that carrying out the analysis over the full domain, where at least one of the functional data is observed, may not be the optimal choice for classification purposes.

C0167: **Amplitude and phase classification of heart conditions using functional data analysis**

Presenter: **Chibueze Ogbonnaya**, University of Nottingham, United Kingdom

Co-authors: Simon Preston, Karthik Bharath, Andrew Wood

A functional data analysis approach to heart defect detection using heart signals recorded by electrocardiograms (ECGs) is proposed. ECGs can be thought of as continuous functions having an amplitude and phase component. However, raw heart signals are usually noisy with artifacts such as baseline wander and there are also issues with arbitrary location and scale when comparing two or more ECGs. To remove these artifacts, we propose amplitude registration models and give closed form solutions for the estimated parameters. For the classification of subjects, we propose to fit mixture Gaussian and cubic spline parametric models (which contain both amplitude and phase components) to the ECG functions. For heart conditions characterised by amplitude changes such as high peaks or inverted curves, classification is done using the estimated amplitude parameters. However, when conditions are characterised by changes in time domain, classification is done using the estimated phase parameters. The predictive accuracy of our proposed approach using leave-one-out cross-validation is 91% for the amplitude classification of myocardial infarction and 96% for the phase classification of cardiomyopathy. Our results compare favourably with state-of-the-art approaches for the classification of ECGs. Additionally, the proposed approach is applicable to the classification of other periodic biosignals.

C0182: **Linear classification for functional data with direct estimation**

Presenter: **Juhyun Park**, Lancaster University, United Kingdom

Co-authors: Jeongyoun Ahn, Yongho Jeon

Functional data are inherently infinite-dimensional and thus dimension reduction is crucial in solving many inverse problems arising in statistical analysis. In that regard, functional PCA has been widely used as a key technique to find an efficient finite dimensional representation. Many regression and clustering solutions are also based on that, as essentially the inverse of the covariance operator is well defined. On the other hand, it is well known that functional classification can achieve a perfect classification, if the infinite-dimensionality is well exploited. This implies that for the purpose of classification, it is not necessarily advantageous to have a well-defined finite-dimensional representation, especially in terms of the inverse of the covariance operator. Based on these observations, we seek an alternative approach to functional classification with a direct estimation method. We specifically consider the problem in linear methods and formulate it as a regularization problem with appropriate penalty. An added advantage of using penalty formulation is the possibility of incorporating some structural constraints in functional data such as sparsity or smoothness as we desire. We study the performance of the new method and develop an efficient algorithm to implement it. Numerical examples are used to compare the performance to existing methods.

C0186: **Functional linear models for energy modelling and disaggregation**

Presenter: **Matteo Fontana**, Politecnico di Milano, Italy

Co-authors: Simone Vantini, Massimo Tavoni

Smart energy meters generate real time, high frequency data which can foster demand management and response of consumers and firms, with potential private and social benefits. However, proper statistical techniques are needed to make sense of this large amount of data and translate them into usable recommendations. Here, we apply Functional Data Analysis (FDA) to identify drivers of residential electricity load curves. We evaluate a real time feedback intervention which involved about 1000 Italian households for a period of three years. Results of the FDA modelling reveal, for the first time, daytime-indexed patterns of residential electricity consumption which depend on the ownership of specific clusters of electrical appliances and an overall reduction of consumption after the introduction of real time feedback, unrelated to appliance ownership characteristics.

Sunday 02.09.2018 16:00 - 19:00 Parallel Session P – CRONOSFDA2018

CI050 Room C413 SUMMER SCHOOL SESSION XII **Chair: Fang Yao**

C0198: **Functional data analysis and related topics**

Presenter: **Fang Yao**, Peking University, University of Toronto, China

Functional data analysis (FDA) has received substantial attention in recent years, with applications arising from various disciplines, such as engineering, public health, finance etc. In general, the FDA approaches focus on nonparametric underlying models that assume the data are observed from realizations of stochastic processes satisfying some regularity conditions. The estimation and inference procedures usually do not depend on just a finite number of parameters, which contrasts with parametric models, and exploit techniques such as smoothing methods, dimension reduction that allow data to speak for themselves. This tutorial will cover general ideas in functional data analysis, such as functional principal component analysis, basis representation models, functional linear regression as well as more flexible regression type models, and so on. Some basic computing and data analysis using R and/or Matlab will be also introduced.

Authors Index